No Waste

This book is dedicated to Mike Malina of Energy Technology Associates and Val Lowman OBE of BeOnsite. Mike introduced me to this industry and Val showed me it has a heart.

No Waste

Managing Sustainability in Construction

ULY MA

GOWER

Published by
Gower Publishing Limited
Wey Court East
Union Road
Farnham
Surrey, GU9 7PT
England

Ashgate Publishing Company
Suite 420
101 Cherry Street
Burlington,
VT 05401-4405
USA

www.gowerpublishing.com

British Library Cataloguing in Publication Data
Ma, Uly.
No waste : managing sustainability in construction.
1. Sustainable construction.
I. Title
690′.0286-dc22

ISBN: 978-0-566-08803-2 (hbk)
ISBN: 978-1-4094-3646-1 (ebk)

Library of Congress Cataloging-in-Publication Data
Ma, Uly.
No waste : managing sustainability in construction / Uly Ma.
p. cm.
Includes bibliographical references and index.
ISBN 978-0-566-08803-2 (hbk) -- ISBN 978-1-4094-3646-1
(ebook) 1. Sustainable construction. 2. Sustainable
construction--Management. I. Title.
TH880.M2 2011
690--dc23

2011017154

MIX
Paper from responsible sources
FSC® C018575
FSC www.fsc.org

Printed and bound in Great Britain by the
MPG Books Group, UK

Contents

List of Figures

List of Tables

Preface

Construction is about building things that stay put. Sustainability is about changing to evolve.

Construction has been around since way, way back. While sustainability is the latest corporate fashion. Mixing the two sounds ambitious, the irresistible force meeting the unmovable object. The encounter is both exciting and painful as well as providing opportunities and threats.

This book is about how to manage this encounter. It is about sustainability in construction as a management activity. It is not a technical guide for the sustainability specialist nor is it about legislation, regulations, official guidance, construction technologies, building materials or any of the environmental/green technologies. I decided against any technical coverage because by the time you read this, any new wonder technology I described is likely to be obsolete and if you wanted to know about construction technologies, just read the trade press every week. The emphasis of this book is about managed sustainability just like any other business activities and to manage it effectively and efficiently.

The other reason I decided to focus on managing sustainability is that our industry has already shown that we can build to various sustainable requirements, such as BREEAM, LEED or Code for Sustainable Homes.[1] The fact that we are mostly doing the actual building activities in a not too sustainable way is what needs to be addressed.

For the manager, this book helps put you into the right frame to 'do sustainability'; for the sustainability specialist, this book provides a platform to engage the entire workforce towards a more sustainable way of working. This book is therefore aimed at addressing the needs of the 'rest of us'.

1 Various specifications for sustainable buildings for UK commercial, USA and UK domestic markets respectively.

A further clarification: I am not a 'sustainable developmentalist' or a 'sustainable developer', or even a 'sustainabilitist'.[2] Instead, I am a 'continuous improver': I believe we all have the capacity to improve on our achievements. This book represents my personal outlook on sustainability in construction. It gives a snapshot of construction and sustainability in the UK as the industry moves from the good times through a recession to a sluggish recovery. It also has a bundle of tools to help you enhance your own and your company's effectiveness. But most of all, it is an introduction to those who wanted to get everyone at work to 'do sustainability'.

Getting everyone to do sustainability means the book has to be practical and easy to use. So I did not write it as an academic tome or a text book. My aim is that you will find the ideas and tools in this book interesting and useful so you will use them tomorrow at work. Work in construction is anything from design to analysis to building to keeping client, partners, suppliers, neighbours, law makers, do-gooders and everybody else content.

This is not a simple or straightforward subject. Had I known this when I first met and discussed the book project with Jonathan Norman, my publisher at Gower Publishing, I might never have shown up at that meeting! The subject is huge, it has enormous depth as well as breadth, and that is only considering the things I come into contact with in construction. It has more topics and interesting stories than I can cover in several lifetimes. A friend told me that he knows of over 70 trades in the industry, and that does not include trades and professions outside the construction site.

So the book is limited to what I can cover realistically while still making a living as a management consultant. It also has a London focus, for this is where I live and do most of my work.

As I was preparing to write this book in December 2006, I met with Kristina Smith, then editor of *Construction Manager*, a UK construction trade journal. She challenged me on whether there is really an interest in sustainability in construction. She also told me a reader wrote to her claiming that sustainability is basically 'emotional claptrap'. That got me thinking, is it claptrap? If that is what some people are thinking, then there is a need for information to help them decide and make sustainability happen.

2 None of these are real words, I do not know what is the one-word noun for the person who does sustainability or sustainable development, and so far, neither does the *Oxford English Dictionary*!

A colleague asked me whether this book will be a hard book and that got me thinking. In a *Fortune* magazine article[3] on 'excellence', it was reported that 'research now shows that the lack of natural talent is irrelevant to great success. The secret … is ... painful and demanding practice and hard work.'

I am not into pain and I do not believe in putting other people in pain either. Furthermore, I always believe that there is no need to do anything the hard way if there is an easier way available. (Easy is also likely to be more energy and resource efficient.) If sustainability is made too painful, people will not adopt it willingly, if it is too difficult, people will not understand it and if it is too demanding, people will avoid it. I wrote this book to show that sustainability need not be painful, demanding or particularly difficult. You need to be dedicated, committed (and have a bit of budget) to make it work, but that is true about more or less everything at work and besides, you know that anyway.

3 *Fortune Europe* edition, 6 November 2006.

About the Book

This book is made up of two main components: the book itself and an accompanying CD.

The book itself has two themes: talking about sustainability and doing sustainability. The talking part contains some background on sustainability in construction. I will explore sustainability and its drivers, how the need for money can spoil everything, whether to go forwards or backwards. Although bricks and mortar are not covered in this book, people are, the social aspects of sustainability in construction are investigated as are the skills debate and communications. Marketing sustainability and the ability to laugh at ourselves are examined alongside making sustainability happen by being pragmatic and professional.

Following a definition of sustainability comes the definition of anti-sustainability. By focusing on how to get rid of the negatives, I hope to show you that sustainability can be achieved by everyone easily, cheaply and in a way that engages the entire construction team.

Having said that this book has two themes: talking and doing. I also introduce an in-between theme, a combination of both talking and doing: this is the government policies and agenda. Much of it is talk, of course, but the policies also compel action, so there is some doing too. I choose three elements from the last few years as my examples: the Codes for Sustainable Homes; the Site Waste Management Plans; and the Strategy for Sustainable Construction.

After the talking is done, we get to the doing. This part of the book covers gaining a corporate consensus on what should be done, through assessing current activities to improving existing processes and transforming the company. These chapters are packed with various tools, from a scorecard for social sustainability to tools that help you assess issues that are critical to sustainability.

The book is closed with a discussion on the Rethinking Construction Report by Sir John Egan and how this could look if it is reissued now.

The accompanying CD is not designed to be a coaster for your mug of tea or coffee. It is intended to be used in conjunction with the book. It contains all the graphics from the book in full colour and sorted by chapter. Every tool mentioned in the book is described together with usage guidance, worksheets and templates where appropriate. Since it is likely that you will need to convince some people to change their minds about making sustainability happen, I have included additional graphics from various topics in the book for you to use in your presentations. There are also a number of generic icons which are suitable for all kinds of computer graphics applications.

All of this is wrapped up in a web interface, which should open readily in your browser.

A Word About the References

I know many books need footnotes and endnotes to show that the author actually read something and did not make up the entire contents. I provide some notes and comments here and there to entertain and inform. However, some of these references are websites and the links (if provided) are good until the webmasters decide to move the particular page.

A Book Called *No Waste*

The name of the book is *No Waste*. Those of you who worked with me may remember a project of mine called NoWaste. This book is not about my project, although it is described in an appendix. I called this book *No Waste* because that is the message I want to pass on.

Acknowledgements

Just like any construction project, getting there requires people working together and supporting each other. This book is in your hands only because of lots of people giving me encouragements, support, criticism (constructive or otherwise), ideas, knowledge and food. The support has been generous and automatic, no sooner than I mentioned I am writing a book than the answer comes: how can we help?

I want to thank especially my publisher, Jonathan Norman and Mike Brooks, my editor at Gower Publishing. If you guys weren't there, neither would this book.

Some of my industry colleagues have moved on to new endeavours since they spoke to me, I have recorded their affiliation (alphabetically) at the time of our discussions.

Thank you to Peter Head, Andrés Luque and Jeff Teerlinck from Arup; John Ruffle of Allenbuild; and Colin Read from Balfour Beatty Workplace. Big thanks to Andrew Kinsey for organising a workshop at Bovis Lend Lease for me as well as championing the NoWaste project at the large construction project in Stratford among many other bits of help. Thanks to Nick Clarke, Paul Edwards, Carmen Garbe, Sam Hall, Val Lowman, Kirstin McIntosh, Steve Wheeler and many others from Bovis Lend Lease.

And thank you to Katherine Adams, Mindy Hadi, Gilli Hobbs, John Musset and other at BRE. A special thanks to Claudine Blamey at British Land who let me bother her many times. Many thanks to Phil Clarke of *Building Magazine*/ UBM and to colleagues at Canary Wharf Contractors: Bianca Stendtke and Seon Waterman. I have learnt lots from Carbon Trust's David Vincent and from Carillion's Nick Drake, Stuart Mee and Louise Perry. Charles Rich of Charles Rich Consultancy provided me with unique insights on construction, of which

I am very grateful for. Thanks to Susan Steensma from CIRIA; Jon de Souza and Zoe Le Grand Constructing Excellence who readily cheered me on. I do appreciate your support!

My hard nosed colleagues, Chris Massie, Gerry McCarry and Mark Townley at Clipfine Construction Support Services showed me how different contractors tackle sustainability in their own unique ways. Anne Hall and Guy Hazlehurst explained to me their efforts in laying down a lot of ground work on sustainability at ConstructionSkills; thanks also to Pat Bowen, Alistair Colin and Hayley Merry at the Future Skills team at CSkills for their encouragements; and thanks to Costain's Amy Baldry, Darren Dobson, Peter Fisher and Phil May for sharing their thoughts and ideas.

David Hughes and Clive Young at DTI/BERR/BIS were always patient and supportive whenever I turned to them for advice. This brings me to the patience of my colleagues. My obtuse questions and lack of awareness only brought a more detailed response, especially from busy people like Murray Coleman of Bovis Lend Lease; Glen Irwin of Foreman Roberts; George Stowell of George Stowell Architects Urban Design; and Peter Rogers of Stanhope. I am very grateful for both your time and your ideas.

Thank you also all the colleagues who shared their insights with me: John Hutton of Edmund Nuttall (now Royal BAM Group); Mike Malina of Energy Solution Associates; Roy Casey of Foreman Roberts; Rob Harrison from Fosters + Partners; Steve Downing of Henley Management College; David Hopwood and Katerina Maluenda of Kew Design/Build who showed me how an SME design bureau managed sustainability issues; David Farebrother and Alex Herd-Smith of Land Securities; Stuart Crane, Jim, Gaffney and Stephen Thomson of Laing O'Rourke; Ali Mafi of Lean Thinking; colleagues Trudy Harpham and Colston Sanger from London South Bank University; Samantha Heath of LSX; Phil Henry of Polypipe; Deltcho Vitchev at RFI who gave me the global big picture about project investment; fellow Lean practitioners Richard O'Conner of 6ix; Stuart Jessop of Shepherd Construction; Greg Chant-Hall and Simon Weller from Skanska; Andrew Mellor of PRP; Sheppard Robson's Andrew Bowles and Alan Shingler; Michael Kohn of Slider Studio; Warren Vaughan at McCarthy & Stone; Neil Johnson from Taylor Woodrow Building; Stuart Green of University of Reading; Wilfried Wang from University of Texas; and Thiru Moolan and Rhian Williams of Wandsworth Council.

A special group of long-time colleagues provided ideas, content and opportunities to debate various topics. Thanks to: Sij Dhanju of the Highways Agency; Lynn Williams of Lunes Training; Anastasia Marinopoulou of Progress Through; Julia Thrul of VGL; Peter Willmott of Willmott Solutions; Isabelle Beaumont from Workplace Futures; and Andrew Pears of Kotuku who not only shared his ideas with me, but also fed me sandwiches from the Cafe Van!

Thanks also to a partner who has worked together with me since joining the construction sector, my colleagues at UCATT: Pat Dowling; Shirley Foley; Jeff Hopewell; and Brian Rye. A special thanks to Jen Renz of CH2M Hill in California and to Dimitris Paterakis in Greece.

Very special thanks must go to Rebecca Lovelace of Circle Three Consulting, who not only explained things to me, also inspired both a chapter and a checklist to the book; and to Mike Harris, a teacher, colleague and friend who is always patient in explaining things to me.

Finally without my wife Karen the task would have been much, much harder – thank you.

Introduction

On 2 July 2008, I gave a talk on the links between sustainability and good management at a CIRIA[1] seminar in London. Despite the interesting content of my talk and my brilliant delivery(!), I did not manage to fire up or excite the delegates; because on that day, the construction industry in the UK fell off a cliff. On that day Taylor Wimpey's share price dropped 52 per cent and 900 people expected to lose their jobs. It was also reported on that day that the big Stratford Project may lose a quarter of the planned number of apartments due to the credit crunch.

Suddenly the emphasis shifted and all my friends in the construction industry had something new on their minds – will we survive this? Unlike other recessions and economic shocks, this one has deeper links to construction. It started with the crisis from the 'subprime' mortgages in the USA, resulting in the home building sector there getting hammered first. Then the banks realised that they had a crisis, since nobody knew who owed the mortgages which may never be repaid. When the banks decided that they did not want to lend to other banks, which might be carrying loads of duff mortgages, every bank in the world got hammered by the resulting credit crunch.

When banks have no money to lend, even people who can still afford mortgages cannot get mortgages anymore. Since nobody is buying new homes, the UK home builders got hammered next. But when banks have no money to lend, then developers cannot get loans to develop commercial property, and so the commercial builders get hammered too. And then the recession hit.

Now a new UK government has decided that we cannot run up endless deficits and so cuts are made to major initiatives, such as the Building Schools for the Future initiative. Although in mid-2010, the UK had a mini recovery led

1 Also known as the Construction Industry Research and Information Association, a membership organisation in the UK construction industry.

by the house building sector – are the good times coming back? When I first started writing this book in January 2007, the world was very different. Then, everyone was telling me that sustainability was on the way in and its progress was unstoppable. The focus was on how we could deliver sustainability – our government was developing all kinds of policies to make sustainability a reality. By the autumn of 2008, the new Sustainability Strategy for Construction (born on 23 June 2008) seemed eclipsed: the discussions, applause and boos were much quieter than expected. Has the inexorable progress of 'sustainability' met its match?

This is likely to be the first real crisis for sustainability since its birth sometime in the early 1990s. Sure, there have been crises aplenty since 1990, but none as bad as the current one. Everyone and every project is affected and recovery seems a long way off especially if global recession is going to be around for a while.

One of the affected activities was the original introduction to this book. It was written in better times with a lighter heart. But as we are in an industry that works hard at reusing and recycling, have no fear, many of those words will find their way back somewhere in this book: our aim is no waste!

When sustainability collided with the recession in 2008, the construction sector gasped and thought: what happens next? Before many managed to formulate a solution, we are in a sluggish recovery. For many construction companies, the answers can be a multiple choice (see Figure I.1) like the CSCS Touch Screen Test:[2]

0 'Do nothing'.

1 'Do the minimum' – slash and burn: fire people; clamp down on spending; squeeze subcontractors; and panic.

2 'Full speed ahead' – pretend nothing is happening and forge ahead regardless.

3 'Assess, review and improve the business.'

4 'Use this opportunity to transform our business.'

2 The CSCS Touch Screen Test is a basic health and safety test for the construction industry developed by the Construction Skills Certification Scheme.

Unlike the CSCS test, you mess up this one and your company goes down the drain. And unlike computer games, there is no 'undo' button or 'three lives' either. You generally get one shot at this and you had better get it right!

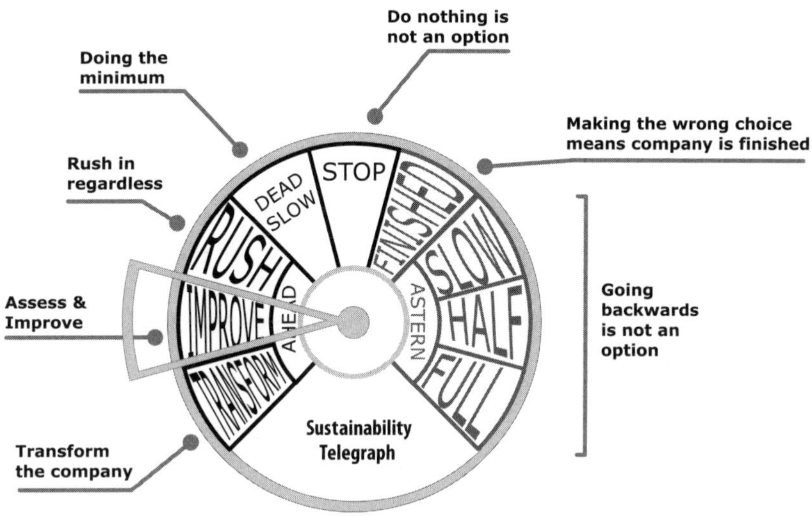

Figure I.1 The sustainability engine room telegraph

0. Do Nothing

This is not really a solution because legislation is in place to make us do something, but it is possible to do nothing. If you choose this option, then chances are we will see you in court (either the bankruptcy or the regular variety).

1. Do the Minimum

This sounds like a terrible option, but many companies are considering it. This is seen as a nearly no-risk option because there will be hardly any new ventures (well, even cautiously exploring some public sector work is not too likely after the 2010 cuts). Since many of our competitors are also doing this, there is little to lose. Nevertheless, getting rid of people means losing talent and squeezing subcontractors means losing goodwill. Neither are particularly crucial when nothing is happening but they will both start to matter when the good times come round again (unless, of course, a major contractor buys your company before then).

What happens is that companies do the least they can get away with under the law and avoid spending any more money than they absolutely have to.

2. Full Speed Ahead

Clearly, very few companies will choose the 'pretend nothing is happening' option although there may be some companies which are so committed to sustainability that they will not drop their efforts even during a credit crunch and recession. Generally though, it is more likely that some dedicated individuals may adopt this – perhaps a group of sustainability specialists who are fully committed to the cause. In the current economic climate, there may be a fear that their jobs are at risk and so these professionals are making sure the company boards are not moving away from the cause since it can also mean saving their own jobs.

Some sustainability specialists will be suggesting pushing this agenda in a difficult time as a means to gain and extend any competitive advantage. Although sustainability in its environmental management perspective is unlikely to suffer because much of it is backed by legislation, the do-good variety of social sustainability is in a much tougher place. Even if no company will (publicly) claim to turn its back on sustainability, many will be looking, possibly in vain, for third-party funding and others may scale back their activities considerably. So an investment here may pay off. But what is it about sustainability that makes it a riskier portfolio during a recession? It is the same as that affecting training, welfare and research: it costs money but delivers delayed outcomes that are not immediately or easily measurable in cash terms.

Before you start screaming, yes, I did say: sustainability costs money. This is a very contentious subject and I have been shouted at (but not down) by sustainability specialists for saying it. My answer remains the same: in terms of the 'sustainability' that these colleagues are talking about, someone has to certify virtually every product and service for it to be accepted as 'sustainable'. Certification itself requires some further certification and we end up paying the salary of an entire industry to assure us that the sustainable products and services we purchased are truly sustainable.[3] Then there is compliance. It takes time, effort and resources to deliver legal compliance and that also costs money. We also need to include the salaries of the technical specialists as well as their monthly mobile phone bills to the increased costs too!

Compared to the 'non-sustainable' variety of products and services, which carries none of this additional costs, there is no economic way that a more sustainable product or service can actually costs less. There is simply more value-added through the certification and assurance and as we know, more value-added always costs more money. Money appears to be the root of the problem, which lead us to Options 3 and 4.

A recession means tough times, but it is also a time where companies can push through difficult changes. Individuals and teams are fully aware that unless they are willing to change, it could be the end for everyone. Therefore it is likely that there will be fewer objections and complaints and less resistance to change. With our collective backs to the wall, the emotional environment can help to get rid of one of the biggest headaches in introducing change – 'we have always done it this way'. This excuse is no longer valid and people will just have get on with doing things differently instead of protesting. (Rest assured, people will still moan about it.)

Options 3 and 4 are linked in that Option 3 allows you to remain competitive and Option 4 helps you to gain extra competitiveness.

3. Assess, Review and Improve the Business

Use the recession to take stock, think about the options for the company, where it wants to be, where it needs to be, where it can be and where it is now. Then set out to improve the way things are done in the company and in doing so, you will be able to reduce waste and save money.

3 The Forest Stewardship Council (FSC) website is a good example – it is a website with lots of useful data on timber products. However, somewhere, someone has to pay for all the data gathering and organising activities.

4. Use this Opportunity to Transform our Business

This is an opportunity to explore how we can transform our company from 'business as before' to 'a sustainable construction business' or business as it should be. As a consultant, you may expect me to recommend this option always, since that is how consultants earn mega fees! But transformation is not always the first or the best answer. While transformation is a realistic option to gain a sustainable competitive advantage, it does not happen just because the boss declares it on the corporate mission statement. It can only happen when the entire company share the determination for change and when people are trained to 'do change'. Otherwise, it may not live up to expectations.

People and companies become more ready for transformation after doing Option 3 – assess, review and improve the way work is carried out.

This book will focus on Options 3 and then 4. From Chapter 11 onwards, it will start with ways to explore your company, how it currently works and analyse what can be improved. A successful and continuing improvement programme forms a good foundation for transformation.

The engine room telegraph (shown in Figure I.1) is a very apt metaphor. What if there is a disconnection between the corporate decision makers (the bridge) and the people delivering the value (the engine room)? See Figure I.2. The corporate message can be garbled or misunderstood. Or it can be delayed because the workforce have no idea what to do? It is also possible that the engine room does not buy in to the corporate solution and as a result, there is no action. In a lot of companies nowadays, the sustainability talk is very popular since it is fashionable and cool to be sustainable but there is some doubt whether this message is really carried across to the people doing the work that delivers the sustainability results.

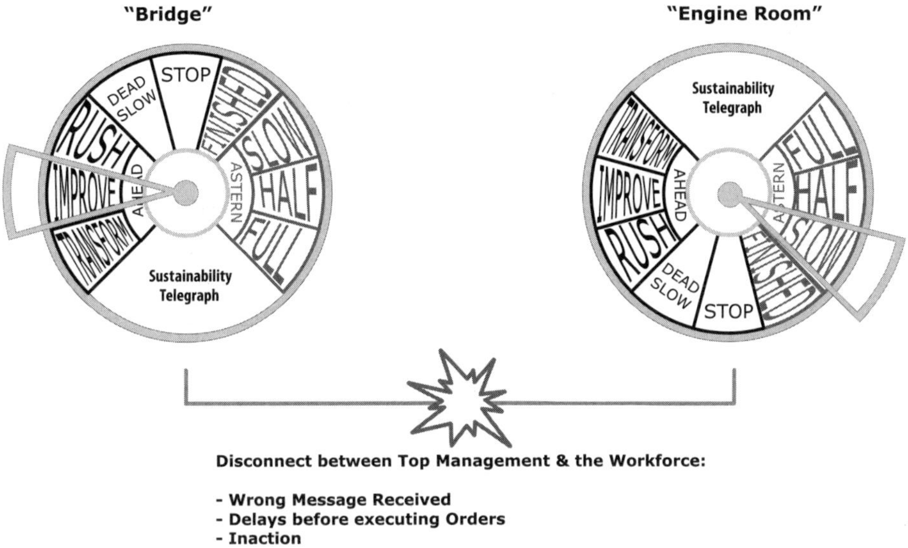

Figure I.2 Disconnected communication

But before we start on all that, let us gather some background information. There are some questions that need to be explored, such as:

- What is sustainability?
- Are there different types of sustainability?
- How did all this come about?
- Is being green the same as being sustainable?
- What has all this got to do with construction?
- Are we really short on skills or are we actually short on 'green skills'?
- Just what are 'green skills' anyway?
- Can construction really be sustainable or is it just marketing hype (or 'emotional claptrap')?

- Just how likely are we to go and do sustainability now with the industry in a depressed state because of the global recession or the tepid recovery?

Besides the questions, we also need to understand some truths about sustainability:

- It is all things to everyone and therefore it is hard to get it right.
- It belongs to everyone so it is difficult to tell people what to do.
- It is both emotional and practical, so it is not easily managed or regulated.
- We all want it – whatever it is, but we'd rather our neighbours try it first!

PART 1
Talking About Sustainability

1

Sustainability and All That

What is Sustainability?

Murray Coleman, former CEO of Bovis Lend Lease, told me that one problem in raising sustainability awareness is that sustainability has many different definitions and aspects. Because sustainability has different meanings to different people, then with over 1 million people working in UK's construction sector, we can estimate that there are probably 1 million different definitions of sustainability or so! So what? Nobody really worries about what construction workers think, do they? Well, Murray does and so do many in our industry. Because many of these definitions stir the emotions, then it can lead to contentious discussions, shouting matches, protests and other disagreements. In other words, whatever you are doing in sustainability, there is a good chance that someone else reckons you are doing it wrong! This does not help our industry going forward.

However, to many companies, sustainability is a much simpler issue. Sustainability, you see, is the new 'black'. In the fashion world, every season has a new 'black' colour or style that rapidly becomes a 'must-have' item. Sustainability therefore becomes a must-have element in every organisation's management objectives. Does this matter? Many of my colleagues feel that sustainability should be taken more seriously, but according to Wilfried Wang, architect and professor at the University of Texas, 'sustainability is a cultural problem, embedded in unsuitable lifestyles'.[1] In other words, maybe having sustainability as the new black is one way to get people to change their lifestyle.

Being the new black has some advantages and disadvantages. The positives are that many projects that 'do-good' instead of making a profit are also approved by companies. It also opens up corporate thinking from a narrow focus on money to broader issues. However, it also introduced us to political

1 Wang said this in the *Fast Company* magazine's June 2007 issue.

correctness, greenwash[2] and a host of other 'emotional claptrap'. Most of all, we also realised that sustainability is not just the new black, the desirable must-have for this season, it is also the 600 pound gorilla sitting in the room. You cannot help but notice it, you cannot walk around it and whatever you do will involve it.

What is Driving All This?

Over the past decade, there has been increasing debate about the increases in global temperature. Depending on your view point, climate change is either an oncoming environmental and ecological disaster or a conspiracy among certain politicians, scientists and pressure groups. Of course, if you are sceptical, you may figure it to be something with a bit of both extremes included.

The approach most of us adopt towards climate change depends a lot on what we know or are able to find out. Unfortunately the science of climate studies is a difficult subject,[3] so we rely on what the media and commentators tell us and the media reports in 2010 about scientific integrity on climate change[4] certainly do not help matters. We have at one extreme the authority of the 600-page report[5] and at the other end there is the sensational headlines in the free morning newspaper.

I found reading all this quite difficult (and no, I did not read the whole of the 600-page Stern Review). Neither the technical terms in one nor the breathless prose in the other really informs me or helps me make useful business decisions. In fact, to a pragmatic business professional, it should not really matter whether we believe it or doubt it. The hard facts are that we have to deal with the reality of the situation – sustainability is here to stay. We have to deal with new legislation and manage the expectations of society because sustainability is rapidly becoming a licence to do business.

2 Greenwash is a derogatory term for describing an act that is designed mostly to benefit you or your company but wrapped up as some environmentally responsible activity. It is often used, unkindly, to describe other people's sustainability activities.

3 For a very cynical view of climate change, read the letter by a lawyer to the *Arkansas Democrat Gazette* on 16 April 2007 suggesting that daylight exacerbates global warming (http://www.theregister.co.uk/2007/04/27/global_warming_discovery/).

4 Towards the end of 2009, there were media reports about leaked emails from various senior sustainability figures that suggest some inconvenient facts (not necessarily truths) may be ignored or hidden.

5 The Stern Review: Economics of Climate Change is a massive volume with a 20+ page executive summary, thoughtfully, the tome can be downloaded from the HM Treasury website.

To a professional, deciding about climate change will often include an element of doing something about it. And to do that effectively, we need a clear definition for sustainability that makes sense in the workplace.

Defining Sustainability

So we reckon we all know about sustainability: it is desirable as well as unavoidable. But what is it?

Well, the *Oxford Dictionaries*[6] say:

> *sustainable*
>
> *Able to be maintained at a certain rate or level: sustainable economic growth*
>
> *sustainability*
>
> *A concept that is used to describe community and economic development in terms of meeting the needs of the present without compromising the ability of future generations to meet their needs.*

The second definition is the one most people think about then they consider sustainability. Nevertheless, even though this appears to be the definition the UK government uses, it is actually somewhat short of the whole definition. It has the benefit of being quite easy to remember, but a bit less easy to understand. An earlier definition of sustainability also exists: it is about something that is *enduring* or something that has been around for a while.

There are a number of other commonly used definitions of sustainability, starting with the Brundtland Declaration,[7] which more or less gave birth to sustainability as we now know it.

6 From the *Oxford Dictionary of English* (2nd edition, revised 2005) and the *Oxford – A Dictionary of Environment and Conservation* (2008) respectively.

7 'Our Common Future' – Report of the World Commission on Environment and Development, United Nations A/43/427, 1987.

Sustainable development is development that meets the needs of the present without compromising the ability of future generations to meet their own needs.

This is often supplemented by the Rio Declaration:[8]

The right to development must be fulfilled so as to equitably meet developmental and environmental needs of present and future generations. In order to achieve sustainable development, environmental protection shall constitute an integral part of the development process and cannot be considered in isolation from it.

Other than an increase in wordiness (and split infinitive), the Rio Declaration pinned environmental protection as a clear path to sustainable development. More recent declarations added to the word count, but the focus remains on environmental protection.[9]

With these definitions sustainability can be (nearly) all things to (nearly) all people. This is great for the 'single issue' crowd since it allows them to bang their drums loudly. It also allows legislators to add all manner of unpleasant but necessary rules to further manage (or control or restrict, depending on what you believe) the way we use resources.

Trudy Harpham, professor at London South Bank University, saw sustainability as five components coming together (see Figure 1.1). The erosion of any one of these can cause imbalances in the entire social fabric. For example, if the social capital assets are disrupted, this can lead to poor physical and mental health which will impact on the human capital (low performance); the natural capital (not fully utilised); the financial capital (investments not giving a good return); and the physical capital (assets not optimised).

8 Report of the United Nations Conference on Environment and Development (Rio De Janeiro, 3–14 June 1992).

9 The Kyoto Protocol (1997) an international agreement linked to the United Nations Framework Convention on Climate Change and set binding greenhouse gas targets for 37 countries. Cynics will say that since then geopolitical posturing have diminished the impacts of the more recent follow-ups.

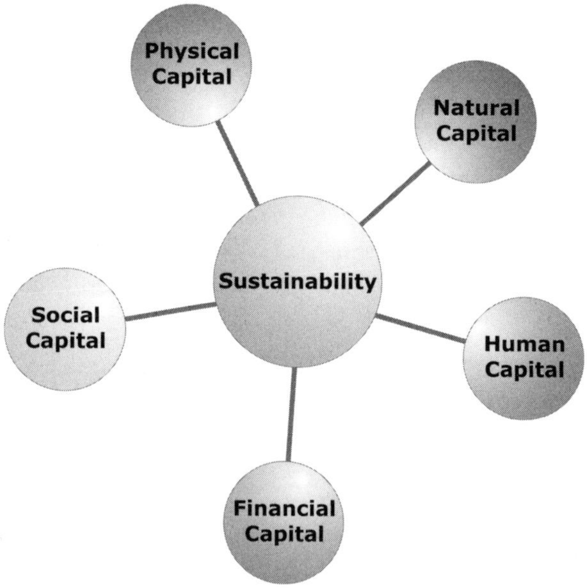

Figure 1.1 Sustainability – a balance of five assets
Source: T. Harpham

Other 'Sustainabilities'

With the construction professional in mind, I propose a somewhat different approach in defining sustainability for this book. First, I believe the emphasis needs to be on the company, organisation or business. This book is not about society as a whole or us as individuals with our personal beliefs, but you and I as business professionals. Therefore, the rather general definitions described above need to be tweaked such that they help you in making better decisions.

The definition of sustainability I suggest for companies is:

> *Sustainability is achieving an effective balance between the financial, social and environmental choices when carrying out business activities.*

The emphases are on 'effective' and 'business activities'. If the business activities are not effective, there is no point doing them at all.

This definition is based on the 'Triple Bottom Line' proposed by John Elkington,[10] (see Figure 1.2) who recommended that companies see sustainability as three elements that impact on the business: the economic; social; and the environmental.[11]

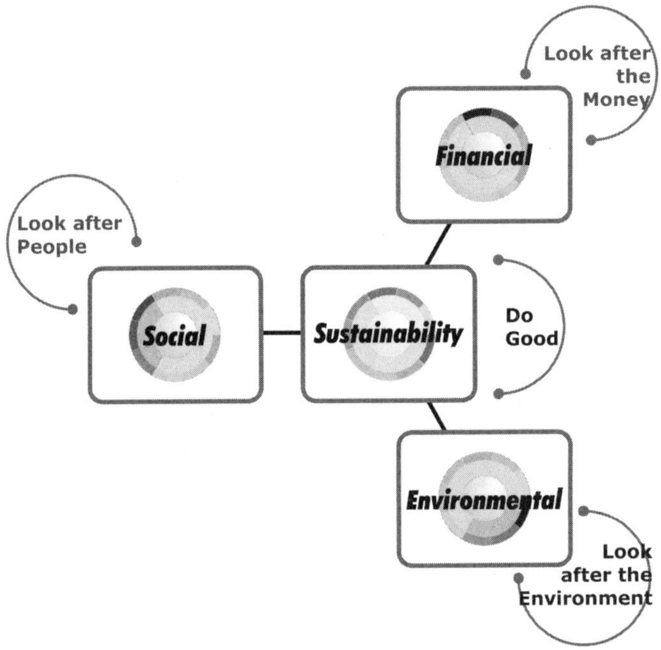

Figure 1.2 The Triple Bottom Line

Let Us Explore This Definition in More Detail

Financial factors are more or less what you expect. We need to manage the financial obligations of the organisation, managing the working capital and managing the cash flow because if we do not, then the company shuts its doors and nothing else, including doing social or environmental good can happen.

10 John Elkington, 'Towards the sustainable corporation', *California Management Review*, 92: 90–100, 1994.

11 I changed from using 'economic' to 'financial' after George Stowell, a London architect, convinced me that 'financial' is the more appropriate word.

I have adapted this model into a nine-element analysis similar to the 'critical to quality' or CTQ model from Six Sigma Quality management. The Critical to Sustainability (CTS)[12] analysis is shown in Figure 1.2 and made up of three clusters as shown in Figure 1.3.

Figure 1.3 Criticality Analysis for Sustainability

Financial Factors

There are three elements: profit; cash flow; and cash in hand. This covers most of the needs in finance. If an activity satisfies all three areas, then it is clearly good for financial sustainability.

12 This sounds like consultant jargon! But it is also likely to be the simplest, most complete and easiest to remember criticality assessment for sustainability you encountered this year!

Social Factors

These are the 'nice stuff' that make the work environment, the work itself and relations with the neighbours more pleasant. These affect three main groups of people: the committed stakeholders comprising the employees, the shareholders and the families of both groups; the involved stakeholders who are the suppliers and customers, and users of our products and services; and the interested – the external or social stakeholders. The committed and involved classifications are best explained by the following quality management metaphor. 'Bacon and eggs' – the chicken is involved but the pig is committed.[13]

If an activity benefits these groups (or a realistic subset of them) then it is 'doing' social sustainability.

Environmental Factors

These can be simplified into one sentence: minimise the nasty impacts of what we do in our jobs.[14] Since everything we do uses energy and 'making' energy causes environmental impact, our aim is to reduce that impact without serious disruptions on the other two aspects of sustainability: finances and people.

The simple three-element grouping for environmental factors is based on pragmatism. Waste reduction and legal compliance are self-explanatory, but I also suggest there are PR opportunities. Since the social stakeholders are as interested in this as the financial partners and backers, then reporting good work becomes a part of maximising the productivity of your activities. Therefore, if an activity reduces wastes and complies with legislation, why not show off and tell the world about it?

Together these elements form a simple criticality analysis for sustainability. This 'criticality analysis' is further discussed in Chapter 17.

13 Roger Palmer and John P. Wilson, Maintaining the energy for commitment to quality, *Training for Quality*, 3 (2): 9–13, 1995.

14 Frances Caincross mentioned this in her book, *Costing the Earth*. She reckoned that everything we do at work (all economic activities) cause some degree of environmental damage. This is a sensible and level-headed way to managing environmental issues. Her 1991 book remains, in my opinion, one of the best early books to examine environmental management in a balanced way.

Is Being Green the Same as Being Sustainable?

Green is a nice colour but it has been hijacked: being green is basically what *I* do all the time while what *you* do is greenwash.

Being green generally means caring for the environment in general but can be stretched to mean being progressive and being 'cool' nowadays. Marketing professionals probably have an entire vocabulary to describe 'greenness' to the public. Being cool, greenness naturally includes all three sustainability elements.

Financially, this covers ethical investment, equitable pay and so on. It can also extend (or heavily stretched) to include income equality and transparency. But mostly it is about investing in processes and systems that minimise damage to the global ecosystem.

In social terms, green is a vague colour. Is it about being nice to people? Is it about being more environmentally responsible? Basically it seems that this is about behaviour change of one sort or another (of course, if you are already nice to people and environmentally responsible, then there is no need to change). However, scaling up from one person to an entire community can smack of social engineering where government policy is steered in one or more specific directions through a mixture of carrot and stick.

Green and the environment are closely linked and probably most familiar, from energy efficient light bulbs to wind generators. At its most simple and basic, it is about not wasting energy. It is also tied into the government mantra of managing waste, the Waste Hierarchy (see Figure 1.4).[15]

To many people, green *is* recycling. Unfortunately, when green is mixed with other ingredients, such as political correctness, a need to show off, as well as some intensive marketing, then the result is often less pleasant: a repressive and strident message that turns many people off. When being green stops being a lot of fun, people start to think about other ways to spend their limited free time. A recession is also a time for people to consider a different aspect of sustainability: their job security and the ability to keep up their financial obligations. If 'green' means a hit on the wallet, then people may become less

15 This has many different origins, I use the DEFRA's definition from its 2002 Waste Strategy as my source.

green. Luckily in the UK, the ever-increasing energy prices[16] can help in making people use less energy, especially in a recession.

Figure 1.4 The Waste Hierarchy

What Has All This Got to Do With Construction?

Andrew Pears of Kotuku (a.k.a. the Cafe Van) once showed me a partial list of legislation that shaped the way we work in the UK construction industry, including:

- Aggregates Levy
- Amendments to the Building Regulations
- Construction (Design and Management) Regulations (CDM)
- Control of Substances Hazardous to Health (COSHH)
- Hazardous Waste Regulations
- Health and Safety at Work Act (HASAWA)

16 The collapse of oil prices in late 2008 was not fully matched by domestic energy prices. The answer, of course, is because electricity and gas suppliers do not supply crude oil to households. So oil price crashes have only a limited effect on your electricity bill (yeah, right).

- IPPC – the Integrated Pollution Prevention and Control Directive
- Landfill Regulations
- Packaging Waste Directive
- Site Waste Management Plans (SWMP)
- Waste Acceptance Criteria
- WEEE – the Waste Electrical and Electronic Equipment Directive.

This is a hefty list and they all have something to do with sustainability (social and environmental). Our industry is heavily regulated because it is an important one. Forget the GDP and employment figures for a minute and consider this: construction is what physically makes our modern economy's infrastructure possible.

Never mind the bankers, the teachers, the doctors or the shop keepers telling you they are critical, just ask yourself this, 'can these people still be so cool if they have to do their work in the open, under all weather conditions?' This is what we do, we design and build the structures and frameworks that allows bankers to trade, teachers to teach, doctors to have hospitals and shop keepers to have a shop to sell stuff. And we do it well. Before the recession, our industry was a major contributor to the UK national economy at around 8 per cent of the annual GDP, employing around 2 million people in Britain. (This means each of us are pulling more than our weight: 6.6 per cent of the working people delivering 8 per cent of the national economy!)[17]

Nevertheless, there is also a shadow to the good bits. Our industry has an annual accident death toll that brings misery to many families and this is not dropping as fast as we all want. In addition, our industry also created over 100 M tonnes of waste in 2007.[18] That's 1,072 tonnes per square mile of UK land area (about 414 tonnes/sq km).

17 Statistics on the UK economy can be found in many locations, these, from before the 2008 recession, were from National Statistics and the BERR – Dept for Business, Enterprise and Regulatory Reform (which pedantic bureaucrat came up with that name?), now known as BIS – Dept for Business, Innovation and Skills.

18 This figure is from the Draft Sustainable Construction strategy, BIS, 2007.

But what has all this got to do with sustainability? Since we are not perfect, just about anything we do can create some waste. And as we have to do it in the open air and under difficult weather conditions – no clean rooms like the people making computer chips or doing surgery, is it surprising that we make so much waste?

Sustainability is what it says on the box: s-u-s-t-a-i-n – to keep going over a length of time. To keep going, people do all kinds of things: some cut corners; some cut wages; some cut quality; and some end up cutting their own throats. Others do it differently: look after the environment; look after their people; and then look to the government for help when they realise their less scrupulous competitors are not on the same level playing field.

So how should we 'do' sustainability? Would a tick-the-box regime be enough? What is the fuss about sustainability anyway? Does it really matter? Well, it matters and a tick-the-box mentality is not enough. Claudine Blamey, former Corporate Responsibility Executive at British Land, a developer, suggested that we need a 'let's try it' mentality instead and therefore do it properly.

We need to do it properly because it matters. Sustainability matters because it is the new 'black' and it is backed by legislation. It also matters because it is a competitive issue, and it matters because it can also contribute to reducing the impact on the global ecosystem.

But to many people and companies, the most compelling reason is that sustainability is the most fashionable item for this season, even for the recession season and probably the second-favourite management theme after cost-cutting.

Climate Change and Sustainability – Putting it Into Context

In the mid-1990s, we were told that global warming was going to make England as warm as Italy. Back then the optimists thought, 'Hey! This means we can live like people in the Mediterranean!' It is interesting that we thought then it would be good that the UK got to be as warm as Italy. However, some of our super-efficient, timber-framed houses turn out to be a little too warm for the current weather and perhaps global warming now means installing air conditioners.[19]

Instead of experiencing dining by the Mediterranean, we found that global warming might mean having dinners underwater if you live in a major coastal city.[20] (We also call it 'climate change' now because 'global warming' probably sent the wrong message: it sounds too nice and cosy. Besides, climate change can go either way: good or bad and if the scientists and the politicians got it wrong somehow it sounds less embarrassing.)

But the climate change message appears apocalyptic – life as we know it *might* end, *will* end or *must* end. And it is going to happen soon. How soon? Well, if you need to ask, then you probably deserve to eat your dinner underwater. The Stern Review reckons 2050 but the UK Met Office reckons it may start as early as 2009.[21]

There are still voices claiming that climate change and sustainability are conspiracies by nefarious governments or large corporations or worse, by both. However, in some sense, it does not matter whether climate change is real or rumour, nor does it matter how you see it. It does not matter whether you are a fervent disciple or an utter sceptic, it does not matter at all whether you believe yourself a victim or know yourself as a villain. The only fact that really matters is that sustainability is here and it will not go away. Well, it cannot – with so many governments enacting legislation and even more citizens clamouring for more restrictions on business activities.

To the guy who reckoned sustainability is 'emotional claptrap'[22] – what you think no longer matters, instead it is what you do that really matters now for businesses.

19 Not something people in our business will often openly admit to in public hence no one is prepared to stand behind this statement in a quote.

20 Well, that's how I interpreted the Stern Review (somewhat irreverently).

21 This was carried out by the Hadley Centre of the UK Met Office and reported by the Register, a website on technology (www.theregister.com/2007/8/10/climate_model/).

22 See the Preface.

2

Money and Tensions

Tensions Between the Three Aspects of Sustainability

Before the 2008 credit crunch, the main environmental flavours of the month in the UK was about globalisation and carbon footprint.[1] This creates a lot of argument which generally goes something like this:

Table 2.1 The tension between right and good

	Economic/Financial Argument	Green* Argument
Globalisation	Good – outsource to whomever can do it at the lowest price	Bad – think of the carbon footprint
Low cost imports	Good – keeps prices and inflation down	Bad – exploits the workers since they are not paid equivalent UK wages

* Green is used here in the PR, progressive or cool sense, and not in environmental management sense

This is a very simple and simplistic example but illustrates the tensions between what is 'right' and what is 'good'. There is no easy answer since companies need to ensure their own sustainability and will go for the lowest raw material costs so that they can maintain a price advantage over their competitors. Governments deal with this tension by using legislation to control the behaviour of companies and using taxation to control behaviour where legislation is less effective.

1 The carbon footprint is a measure of how much carbon was used to get fulfilment from a product/service. Neither the value-added nor the pleasure of consumption is involved in the calculation. As a strictly environmental measure, it works, except people either cheat and set ridiculous perimeters ('from our warehouse to you') or have no idea how to set the perimeter ('is it the electricity generator's footprint or ours?'). There are many 'carbon calculators' available on the internet. I do not endorse any of them.

Sustainability is riddled with tensions:

Table 2.2 Tensions between elements of the Triple Bottom Line

Example	Financial	Social	Environmental
Low cost imports	Yes	We lose jobs	Carbon footprint increases
Enhance workplace comfort	'Gulp'	Yes	Carbon footprint increases
Environmentally assured products and services	'Gulp'	Higher costs means lower bonus/dividend	Yes

Some of these tensions come from the generally skewed way businesses look at sustainability. Many see sustainability as an evolution of environmental management, itself coming out of energy management. Therefore the environmental elements tend to become embedded far deeper than the other two elements, especially as legislation also pushes in the same direction. Nevertheless, it is financial stability that forms one of the original definitions of sustainability: great corporations would not be around if they had not managed to sustain themselves over the years. Yet perhaps because finance is run by its own set of specialists (the accountants) and has its own established jargon and culture, it never seemed to have joined up with the more popular modern definitions of sustainability. This is not the case with the social factors of sustainability, as 'being good to people' is not too far from 'being good to the planet'. However, 'being good to money' is more than just an imaginative sidestep, it is a cultural leap across a chasm.

Val Lowman, OBE and director of BeOnsite, Bovis Lend Lease's social sustainability arm, suggests that we should use the tensions to drive positive debate and then make something happen. This is a good idea because it offers the industry an opportunity to look at the way we currently work and maybe see opportunities instead of problems. Val believes that 'we need a mindset transformation in this industry'.

All Well and Good, You Say, But Then What?

The tensions between financial, social and environmental aspects of managing a organisation arise simply because the organisation has to fulfil certain obligations to ensure its survival and often, these obligations clash.

Obligations to Whom?

Organisations have many obligations: the financial ones, such as managing its costs, repaying its loans, rewarding its investors and paying its employees; the social obligations, such as providing a safe, respectful and sustainable employment for its workforce; and the environmental obligations, such as minimising environmental impacts while carrying out business activities.

To deliver on all three without compromise is not going to be easy as we have seen. There will be tensions. For example, some of these can be:

> *Conventional wisdom in construction tells us the lowest cost bid often wins the contract, but the lowest bid may mean a less than equitable wage for the workforce and a less progressive attitude towards environmental impacts.*

Hiring local unemployed people and helping them develop skills will provide a big boost for them personally and also for the local community, but their immediate productivity will be lower than hiring more skilled migrant workers. Furthermore, it may prove easier to exploit migrant workers by paying lower wages and offering poorer working conditions.

Being at the leading edge of environmental technology can bring prestige, but also carry a cost and risk penalty. It will always cost more to be at the leading edge since there is unlikely to be any economies of scale. Deploying new technologies can often be riskier too. Although Peter Head, CBE, director of Sustainability at Arup an engineering consultancy, thinks that there are no perceptible lifetime costs when building in a sustainable way to building in a traditional way. In that respect, the financial risks are lower than perceived by many people. Nevertheless, these tensions do exist and balancing these tensions is *the* challenge facing all businesses in the coming years. And to make a situation even more complicated, balancing these tensions does not necessarily mean equal treatment for all three.

Balancing the Tensions

Over the past few years, the environmental message has been getting louder and louder. The social message, boosted by a thick slice of political correctness, is also getting louder. These changes are fine as long as they deliver a more effective way of doing things. After all, it is a different world now. But sometimes these messages affect the balance between our three aspects of sustainability.

The demands of the financial ('we cannot afford it!'); the social ('people come first!'); and the environmental ('do you want to destroy everything?') put organisations under a significant amount of stress. Your decision in meeting these demands will depend on where your priorities are. Actually it can also depend on where you are physically! Your view about sustainability may vary when you are sitting on your sofa at home as a concerned citizen and when you are at your desk at work and putting a bid together for your organisation. I recall an interesting discussion with a group of construction subcontractors. I was trying to encourage them to adopt a waste reduction programme, but I was told by the subcontractors that, 'we support this in principle and we think it is a good idea. We recycle at home, 'but (waste reduction)' is not in our contract so we won't do it here.'

Why Financial Factors Always Lead in Sustainability

At this point, I will annoy some colleagues and readers by saying that, for a business, the financial aspects of sustainability have to lead the other two. This is simply because if the finances of your organisation are not managed well, your organisation will collapse. When that happens, it does not matter what your social and environmental policies are. If you cannot pay your bills, you will not be able to pay for the social or environmental programmes either.

Boosters for social and environmental aspects of sustainability will always tell you that investment in people or in environmental management will pay off. Of course they will, but often the payback schedule is not the same as the one you get from the bank. The manager at the bank may sympathise, but if you cannot meet your financial obligations, you will not get too far with the bank.[2]

Guy Hazlehurst, when he was a director at UK's ConstructionSkills, showed a diagram where the three aspects of sustainability were distorted in favour of the environmental aspects.[3] He was concerned that too much emphasis was

2 With the credit crunch, nobody is getting much from the banks – so with a prospect that the 'cash is king' days are returning, a need to conserve working capital becomes ever more important!

3 Guy Hazlehurst showed this during a workshop to support the UK government's sustainable construction strategy development in early 2007. The diagram in Figure 2.1 is based on his idea.

placed on environmental matters to the detriment of the other two (see Figure 2.1). This is quite true as much of sustainability practised nowadays came out of environmental management.

Figure 2.1 Imbalance in sustainability

Why is there a distortion towards the environmental aspects? It is not surprising that the environmental elements lead the discussions on sustainability, as Murray Coleman said, 'it is the entry point to sustainability because it is an area people can most readily relate to and there are hard facts to guide discussions'. There are several other very good reasons too: destruction of the global ecosystem is quite evident with the ravages of pollution; destruction of rain forests; and the news that more and more animal species are becoming extinct through human actions. These are highly visible reasons that often appear on the TV screens and in sensational newspaper headlines. Then there is also a significant social reason: environmental awareness is cool and progressive. Just as well since the other side of environmental awareness is the very uncool but pragmatic compliance with a growing list of environmental legislation. Financial issues matter but as Andrew Bowles and Alan Shingler, directors at Sheppard Robson, an architect said, 'once the industry is forced to comply with legislation like the Code for Sustainable Homes, then costs become less relevant because the entire industry is working from a level playing field'.[4]

With a combination of 'cool', the legal requirements and visible damages, environmental management becomes a relatively straightforward item to pin

4 Nevertheless, building to a sustainable design is not exactly the same as building in a sustainable way.

down and define. Once something is pinned down and codified, it then becomes much easier to deal with. (Think of the parallel in finances: a government draws up a new tax rule, then accountants can work out solutions to minimise your tax – for a fee, of course.)

Some of the champions of the environmental aspects of sustainability are also champions in marketing (think pop stars and former politicians). While accountants may wield a lot of power in a company, but they rarely come first in a popularity contest with people who care for the environment or care for other people (or pandas and polar bears). The fact that effective accountants make it possible for both the environmental and social champions to keep going with their messages as well as making sure you get paid every month is not often mentioned. Why that is so may remain one of the smaller mysteries of the universe.

Greg Chant-Hall, at Skanska, understood these problems well. He felt that putting a cost to initiatives, such as waste minimisation, is easy but the value of some social initiatives can be quite subjective. Nevertheless, it is essential to communicate effectively in terms that people can easily understand and relate to. Without the economic metrics, the value of important initiatives can be missed. Once the value is understood, as well as how to achieve it, we can find ourselves in a good position to build commitment and motivation to do more.

3

Going Forwards, Backwards, Sideways or Nowhere?

Welcome to a different world. With a credit crunch, things became a lot less fun, with a recession, doing business became very hard and a slow recovery does not necessarily make things easier either. Furthermore, with the contributions from climate change, the weather probably will not be too good either.

This is the world many of us suddenly woke up to in mid-2008, when the construction sector fell over. Of course, some people have been telling us it has been coming for a long time, but let us not worry about these people. Instead, think about what this means for you, your organisation and your job. Although there are signs that recovery is just around the corner in mid-2009 (or 2010, or 2011), there is little to indicate that we will return to the former ways of working any time soon.

I briefly examined the various options facing a construction sector company in the depressed business environment in the Introduction. From there the book is focused on improving the way we work followed by enabling everyone to participate by transforming the organisation.

But there are many different ways of going forward.

The best way forward in this different world (see Figure 3.1) is the one that fits your organisation. It may be a 'big bang' or it may mean meandering a bit to collect everyone first; or it can mean only taking short steps or even going sideways or backwards before going forward.

Figure 3.1 Going forward is not always straightforward

Going Backwards?

Unfortunately, my experiences on introducing changes, such as Lean construction with some major contractors, showed that before kicking off, there may be a need to go back to basics first, such as establishing good workplace organisation or a discipline of waste reduction. Many of these initiatives, whether environmental management or Lean construction, require a certain level of knowledge within the management and workforce if they are to participate successfully. For these companies to go forward, there is a need for everyone to have the necessary knowledge and this can mean starting at Step 0 or -1 or -2.

Whether going forwards directly, via small or large steps, or even going backwards matters less than making sure everyone is on board and travelling in the same direction. This means getting the entire workforce to participate, it means persuading your suppliers to come on board and getting the support of your customers. This is a crucial element in any success in sustainability. Without this wholehearted support, the risks of something going wrong becomes much higher. These come in several varieties: the visible ones like pollution and other physical blots on the landscape; the semi-visible one of low morale because people are not sure why they are doing these things; and the invisible risks of damages to your corporate and personal reputation.

Despite my gloom about our industry being ready, Peter Head believed that the transformation to sustainability in construction has already started. When Claudine Blamey was at British Land, she told me that more and more companies are adopting a thinking outside the box mentality which is good news.

Reputation Matters

Even though we defined sustainability as managing a combination of financial, social and environmental elements in the business, there are some factors to consider as well and one of these is corporate reputation (see Figure 3.2). A poor reputation may be less important in the era of conglomerates when 'getting the numbers in' was really all that mattered. In our more progressive and more connected era, reputations can be made and destroyed easily by campaigners, blogs, specialist newsletters as well as the traditional media. A company's reputation can be turned upside down simply by its drivers keeping their delivery truck engines running while unloading. The drivers probably care about the environment as much as anyone else but are likely to have no ownership of the corporate sustainability initiative.

Technology further compounds company indiscretions, from software like Google Earth to smartphones that takes videos, everyone has the capability to report poor performance to the whole world – very rapidly. This means if members of your teams are pouring oily sludge down the drains, it may readily end up on YouTube for everyone to see. Reputation takes a lot longer to rebuild than to destroy. For example, the Three Mile Island and Chernobyl incidents shaped the thinking on nuclear energy for decades in the USA and Europe respectively.

Getting suppliers and customers on board is also difficult, but imagine an irrational public response to the following:

> *ABC Architects designed the presidential palace for the blood-thirsty dictator XYZ.*
>
> *RST Construction, a household name, gets its timber from a company that robs indigenous people and rips out virgin rainforest.*
>
> *LMN, a leading developer, tears up picturesque town centres and replaces them with soulless shopping centres.*

Whether it is bad luck or poor judgement, your reputation is going down the pan if this sort of headline greets your company on breakfast TV, the internet, social media or via the red banner tabloids tomorrow morning.

The impact of having a bad reputation can be quite severe in a society like the UK that seems to thrive on bad news and corporate villains. The best case scenario for a first stumble with duff clients and suppliers is to claim that you had bad luck. If it happens again, then the public perception can change quickly from accepting your company having a spell of bad luck to corporate incompetence and a third occasion? I think it will be time for resignation letters from the Board.

The perceptions of a poor reputation go in two directions: internally within your organisation and externally with business partners; and society generally.

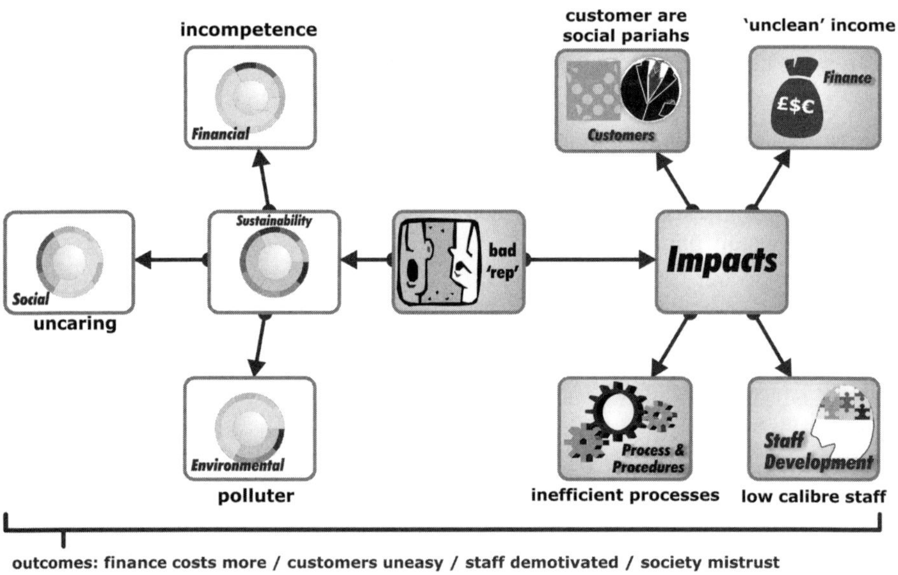

Figure 3.2 Perceptions – the impact of poor reputation

Internally, your disillusioned employees may start to look at the world differently.[1] Your customers are suddenly seen as social pariahs – most people

1 I remember years ago finding out that my then-employer (a conglomerate) also made landmine casings. I felt terrible, ashamed and demotivated. My response was not unusual: recently a few

dislike working for 'nasty people', you will have deflated the motivation of your people if your clients are publicly blacklisted. Since these clients are seen as nasty, then the money they pay your company may be perceived as 'dirty'. This moves the entire bad reputation agenda onto a more personal level: does your income derive from a dubious source?

Bad corporate reputation can also tarnish the personal reputation of the employees. Whether true or not, outsiders may think that if you work for a 'bad' company, then what does this say about you as an employee?[2] Nobody likes this type of stress. Similarly, a bad corporate reputation is not going to help build employee morale unless there is a visible and drastic corporate clear-up.

Essentially, a bad corporate reputation can mean your company being viewed as incompetent in the way it runs its finances, manages people and it also pollutes the environment. Even if these beliefs are totally untrue, you need to make sure every stakeholder, from financial analyst to the neighbours of your project site, knows that they are untrue; and that is hard work. In a depressed business environment, a bad corporate reputation can mean it is even harder to make money or to raise capital, and life becomes a lot less fun.

Construction companies are there to make money and this was evident a few years ago when UK government PFI[3] policies encouraged many traditional construction companies to move into the service sector. Why? Probably because the money is better there. Shareholders and investors demand ever better returns on their investment and companies are driven to grow: growing their turnovers and growing their profits. In the market economy, it seems if your company is not growing, then it is sinking. All this puts a tremendous pressure on making more and more money. Making money is not just critical to keep the company running but companies need to earn 'enough' extra money to pay for the social and environmental aspects of sustainability.

Nowadays corporate reputation can be made or broken by individuals or groups which have no direct connection to the company but have an interest

sustainability specialists were deeply unhappy when they told me that their 'squeaky clean' consultancy also advises on nuclear power.

2 This is familiar ground to people who work for tobacco companies – selling an entirely legal product that just happens to harm its customers.

3 Private Finance Initiatives (PFI) are ways to reduce government spending. Basically, contractors and partners pay for the construction and then get a long contract to manage, maintain and service the facilities afterwards. Like most clever schemes, there were plenty of disasters as well as successes for both sides.

in how the company behaves – these can range from political/social activists to neighbours of a construction project.

Sustainability and Money – a Matter of Balance

One issue that is often present when discussing any 'tension' in sustainability is money. Whether we are talking about tensions between financial, social and environmental issues (see Chapter 2). Money is not always the answer to problems but it does cover up a lot of irritations. What money does, however, is to offer opportunities to ease tensions.

In a recession, money is rather harder to come by and this can lead to a lot more tension in sustainability activities. Unlike other products, there is not a lot of scale economy in a corporate sustainability initiative. You cannot go and get a 'one size fits all' solution because every company and its circumstances are different. Some companies can go forward immediately in doing sustainability while others need to start at a level low enough that everyone can readily get on board.

With sustainability belonging to everyone, we now have a bandwagon that can have more than a million possible definitions (see Figure 3.3). How do we hop aboard a bandwagon like this? Getting everybody on board is hard in this situation because everyone has their own definition of what sustainability should be.

In a construction project made up of many subcontractor, you also get a constantly varying set of 'different definitions' since contractors and their teams rotate in and out. The suggestion of taking a balanced approach towards sustainability for this cast of thousands is hard enough but is made even harder by three other reasons.

The first reason is the vested interest of the individuals involved. This is not about your personal convictions on sustainability but about the way we have been trained professionally. For example, a grounding in accountancy will offer you reassurances that financial controls may be a good approach towards sustainability. Similarly, a knowledge of human resources will encourage deploying people-led techniques to realise the sustainability goals. A builder may reckon it is poor design that causes the problem and a subcontractor may think the welfare policies on site are just not working.

Figure 3.3 One million+ different definitions of sustainability

This is a generalisation but we all have our bias in selecting the best (or our preferred) way to 'do' sustainability. In an organisational context, this can mean that various individuals, departments and functions will all want to pull the organisation down their preferred path. Although this is nothing new and internal fights over corporate directions have always been with us, these conflicts can harden the attitude towards corporate sustainability goals (especially when it is *your* budget that had been cut to accommodate *their* goals).

A second reason for the difficulty in achieving balance is that although everyone knows what sustainability should be, only a few are recognised experts or certified specialists in sustainability. Many sustainability experts come from environmental management and their expertise is well grounded in environmental technology principles or legislative compliance.

In these situations, all those environmental managers are certainly going to apply their skills as effectively as possible to serve their organisations. And since they are experts in environmental management (and technologies) those are likely to be the approaches they will adopt (see Figure 3.4). This then adds an 'official' voice to unbalancing the corporate sustainability policy.

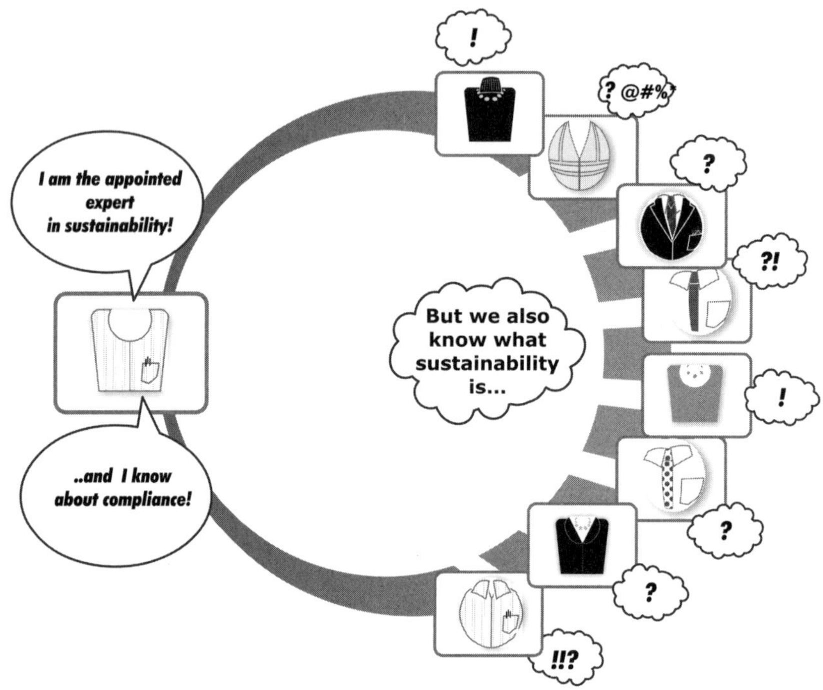

Figure 3.4 ...but we also know what sustainability is

A third reason for the difficulty is that in practice, sustainability is often driven by a combination of legislation, public sentiment and media hysteria. There is some research[4] to show that companies respond to regulations faster than to investor demands, competition or customer needs. So it is not surprising that a good-sized chunk of the consultancy market in UK construction is supported by the need to interpret legislation and provide training for compliance. And guess what, a big element of this set of legislation is environmental legislation. If you do not believe this, just go and visit the website for environment legislation at 'Netregs',[5] run by the UK Environment Agency.

Since most organisations are in business strictly to do business, then compliance with the law is a major obligation. Environmental legislation is a big deal and companies spend considerable effort and time sorting out their responsibilities and liabilities. This is not to say that traditional legislation is no

4 Colin Read of Interserve wrote about this in his MBA Dissertation, London South Bank University, 2007.

5 NetRegs is at (http://www.netregs.gov.uk).

big deal, but their compliance is now part of the corporate furniture and people just get on with it as part of daily work. A more difficult problem is that the amount of environmental legislation just keeps on increasing and that distorts the sustainability balance more and more towards compliance.

Not in Balance

Should our sustainability strategies be balanced? Since everyone in the industry is working under a somewhat unbalanced sustainability portfolio, does it matter at all?

Simply put, an unbalanced set of policies leaves a company more vulnerable in both good and bad times. In the good times, an unbalanced company may not be looking in the right direction for opportunities (see Figure 3.5). During bad times, these companies are still not looking in the right direction for opportunities or threats. Either way, an unbalanced company is less likely to do as well.

Is achieving balance in sustainability that difficult? Actually it is. Besides overcoming the three difficulties described earlier, it also takes leadership and good management.[6] Just before John Elkington came up with the Triple Bottom

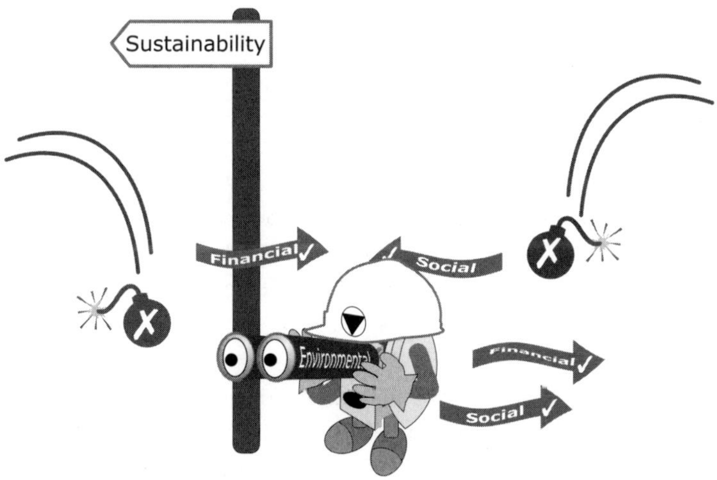

Figure 3.5 Missed opportunities and threats

6 Well, just about doing anything well in an organisation takes good leadership and management.

Line (see Chapter 1), conventional management thinking was tipped out of its comfort zone by the arrival of the Balanced Scorecard (see Figure 3.6).[7] The Balanced Scorecard was designed to encourage companies to shift away from a pure financial focus towards a range of factors that all contributed towards corporate success.

Nowadays with our modern outlook, the Balanced Scorecard actually looked a bit obvious – and possibly politically incorrect (no mention of stakeholders, the environment or legislation). However, in the early 1990s when the corporate focus was 'numbers' (this was the era of the conglomerates), the mere suggestion that processes and the developmental needs of the staff should get equal billing with finance was a new thing to the corporate landscape. Since then, there have been many attempts to update the Balanced Scorecard by bolting on extra elements, such as the environment, legislation and so on.[8]

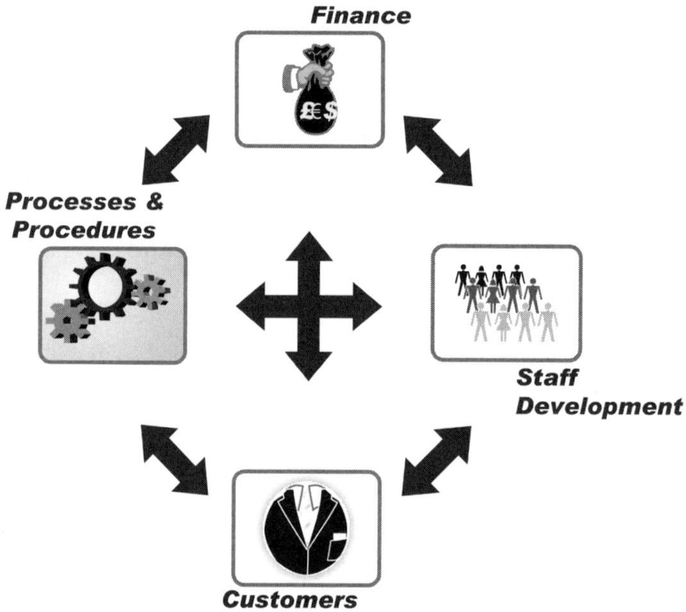

Figure 3.6 The Balanced Scorecard

7 Robert S. Kaplin and David P. Norton, 'Putting the Balanced Scorecard to Work', *Harvard Business Review*, September 1993.

8 Since there is copious room for confusing about the various balanced scorecards, a simple naming convention is suggested, Kaplin and Norton's original is the 'Balanced Scorecard', everyone else's are just called 'balanced scorecard'.

I mentioned earlier that the importance of balance is to allow the organisation to be more flexible so that it can respond better towards good fortune or crisis. This means a company is *ready* to pick up the opportunities as well as to deal with the threats.

'Be Ready' and Sustainability

Although 'be ready' looks like some vague marketing slogan, but it is actually based on sound management practice. Being ready is about the following:

- maintain discipline at work;
- deploy effective methods;
- practise techniques regularly; and
- always able to innovate.

Does this sound unusual? Well, this is how trained professionals approach any situation. Practise effective methods until you gain expertise, rely on your professional discipline when encountering new situations (so no panic decisions) and use your experience (from practise) and competence (from the methods) to innovate a new way of doing things. The four factors of 'being ready' are also pre-requisites for good sustainability performance.

Discipline at work: this means people are moving in the same direction. Makes sustainability easier as it avoid individuals following their own whim as well as any subsequent embarrassments arising from the freelancing.

Effective methods: by having the discipline and following effective working methods, we can reduce quality variations and with consistency, quality can be improved. Improved quality means fewer mistakes and a reduction in cost, reduced impacts to the environment and a happier workforce (nobody likes mistakes) and hey – are we doing sustainability or what!

Practise regularly: means we gain expertise, capability and competence. It also means we can reduce mistakes and errors. The gains in experience and knowledge means we are in a better position to innovate. With innovation and continuous improvement, we can make sustainability happen a lot faster and more effectively.

An ability to innovate: critical for businesses competing in a closed market, such as the UK where most companies offer broadly similar level of capabilities.

Being ready: it is important in good times, but crucial in bad times. However, 'being ready' is an outcome from a corporate commitment. An effective sustainability statement needs to send a message that encourages and enables every employee to get ready. If your sustainability policy is not balanced, then your people will also not have balance in their preparations, and as a result, they will miss some pieces in their readiness. We do not want corporate messages that are finely crafted words with no emotion, and no real substance.

More importantly, since everyone knows what sustainability is, you need to make sure that you have the right message to get across to the every employee including the sceptical ones who reckon you may have got it wrong.

4

People, Construction and Sustainability: The Social Elements

This Chapter of the book is about the social elements of sustainability, beginning with a general introduction and followed by exploring some of the main issues relating to people in construction: people; skills; health and safety; communications; and marketing.

Although construction employs a large number of people in the UK, many work as subcontractors who may not share either the client's or main contractor's vision and culture. There is a risk that both the range of social sustainability activities and the various different interpretations of social sustainability can mean that people may end up doing things differently as well as doing different things. The impacts of not sharing the same set of objectives can be made worse by the large turnover of staff during a construction project as trades move in and out as required. The situation makes effective communications quite a challenge because the same message not only needs to be repeated, but perhaps also adapted for different teams of people. The rapid turnover of staff can lead to some managers feeling detached from their ever-changing workforce. Would you take time out to consider the welfare of people who may leave the project tomorrow and many whom you may never see again?

This diminished enthusiasm is not just about welfare, but also about skills development and health and safety. Perhaps some of it can be attributed to the overwhelming vastness of the social sustainability agenda. Since just about any progressive human resources or public relations activity can become part of a social sustainability campaign, the vastness also leads to an elusiveness –

you are pursuing an ever-expanding and ever-changing subject. Where do you start? (And can you ever stop?)

It does not help when sometimes, social sustainability can appear to be contradictory to conventional business thinking. For example, one element in social sustainability is about the links with the local community. This is where the smaller local building firms are already doing – by virtue of being small and local. According to Stuart Green, professor at Reading University, local firms are more embedded in the community and therefore better poised to support the needs of the local stakeholders. This may be partly because both its clients and workforce come from the same community. Nevertheless, this is the reality for locally-based firms long before sustainability, community investment or corporate responsibility are viewed as something good to do. However, conventional business thinking also suggests that economies of scale can have a significant impact on the costs of a construction business. So does this mean stop growing and stick with the community or grow to deliver a better return on the business investments but become separated from your local links?

Whether being local really encourage a greater sense of corporate responsibility is debatable – think cowboy builders. But what about the community itself? Would a community that comprises like-minded people (in this case being progressive towards sustainability) make an impact in encouraging the industry as well? Michael Kohn of Slider Studio, an architectural practice, is working on a system of housing development called 'Enabled Self Procurement' (or ESP). Michael's view is that residents are interested in how their community is planned and that his proposed system and supporting software will allow them to take part in a coordinated framework where the planning permission, building regulations and the actual construction have all been sorted. This is an example of our industry at its best, listening to the end users – the community.

But does social sustainability mean keeping your focus local? Can a company focus successfully on the needs of local communities while growing to become a regional, then national and finally global construction company? Or does this mean we need to be more flexible in planning our social sustainability actions? All these questions are tough ones to answer because there appears to be no right answer, down to the most fundamental one: is there a definition of social sustainability?

Just What is Social Sustainability?

I put this question to Rebecca Lovelace, director of Circle Three Consulting, a specialist consultancy on community investment in the construction sector. Her answers and ideas helped me shape this chapter and highlighted how different aspects of construction and management activities can work together to form the practice of social sustainability.

Lovelace believed that social sustainability is something much talked about in construction. Although almost everyone shares an agreement that we need to do better in the social aspects of sustainability, it is much harder to define what it actually is.

Looking at the overall picture of sustainability in construction, Lovelace echoed what I mentioned earlier: there is certainly recognition of the economic and social aspects to sustainability, but it has been the 'green' or environmental issues that has been the main driver for sustainability in general. The need for legal compliance on environmental issues and the provision of a response to media pressures has meant we have the skewed system described by Guy Hazlehurst in Chapter 2.

A cynical approach to how construction professionals and managers see sustainability currently could therefore be summed up as:

- is this project financially sustainable? (or: can we afford to build it and will it make money?);
- are we meeting environmental legislation requirements? (or: are we going end up in court?); and
- are there enough skilled workers for the lifecycle of the project? (or: will we need to pay more to get the skills we need?).

Nevertheless, at least people are thinking about sustainability, however cynical their thoughts may be. Still, in my discussions with Rebecca Lovelace, she felt that the social element – or the 'people side of things' – is still placed on a lower priority than the environmental and economic elements. Fair enough, finance is important since we need to make sure that we do not go out of business. Environmental sustainability has teeth because of the many laws –

nobody wants to damage their hard-earned reputation through going to court or paying fines and risk going out of business.

But as she asks, is it simply a case that no one quite 'gets' social sustainability because it is such a massive issue with no real framework within which a clear definition exists? How can construction companies justify spending time and effort to build stronger and healthier societies when there's a building contract to complete? Is it really down to us in construction to build a better and stronger society? Are we supposed to be building not just the physical part of society but also the social bits? Should this be something the government is responsible for? Is this not what our taxes are for? Or should these be carried out by us in our role as volunteer citizens as suggested by the UK government?

These questions and attitudes are on many people's mind. We just do not voice them because they cause embarrassment. It is embarrassing because we seem to be saying that the government is not up to it, embarrassing also because it can also say that we do not care about other people. But wait a minute, are we not in a market capitalism society where only the fittest survive? Yes to the market capitalism but even that is evolving. We are now in a more compassionate society, market capitalism or not. Once again, our 'business as before' philosophy needs to evolve and adapt so we can thrive in the current social climate. It may mean the cost of doing business is higher than before. But with both social pressures and legislation, there are ample reason to change and deliver a more effective business formula that encompasses the social needs in construction.

In our discussions, Lovelace further asserted that we in the industry are already doing a lot of progressive activities but it is just that we have not put it all together and called it social sustainability. This is somewhat compounded by not having a set of real industry-wide benchmarking/reporting tools. While this is an on-going challenge for some companies, many others are already doing things that are often considered to be part of Corporate Social Responsibility or Community Investment, or under the banner of Human Resources, Health and Safety or Learning and Development. Lovelace suggested that if these activities are grouped together there is more potential for industry-wide benchmarking as represented in the global level guidance, such as the UN Global Compact[1] and the International Finance Corporation's Equator Principles.[2] She commended

1 You can look this up at the United Nations Global Compact website (www.unglobalcompact.org).
2 More on the Equator Principles (www.equator-principles.com).

The London Benchmarking Group as a good example of companies working together to measure Corporate Community Investment.[3] The problem, perhaps, is that although social sustainability is present in most companies but often these are used as elements of competitive edge to distinguish one company from another.[4]

The Global Compact

The UN Global Compact is about businesses behaving in a more responsible fashion. Unfortunately, it is both vague and somewhat bureaucratic in its original wording. I have prepared the following abbreviation stripped of much of the original jargon – not a lot can be done, however, for the vagueness. The original version can be downloaded from the UN Global Compact website:

- Businesses supporting and respecting internationally proclaimed human rights.
- Business to make sure that they are not part of human rights abuses.
- Businesses supporting the freedom to form and belong to trades unions.
- Forced and compulsory labour should be eliminated.
- End child labour.
- No discrimination in employment and occupation.
- Businesses should take greater care when dealing with sensitive environmental issues.
- Businesses should establish activities to promote greater environmental responsibility.

3 Find out more at (www.lbg-online.net).

4 If you look at the corporate websites, just about every major company has a section on social sustainability. They all say more or less the same thing, even though they all say it rather differently.

- Businesses should support the development and spread the use of environmentally friendly technologies.

- Businesses should work against corruption in all its forms including bribery and extortion.

Although at first glance these may seem to be stating the obvious, but it is clear that even in advanced rich countries like the UK, we read about these types of problems. Poor health and safety practices by a contractor is an abuse of the human rights and dignity of the workforce; blacklisting workers is a form of discriminatory practice; similarly, the Office of Fair Trading's investigation (2007 to 2009) into price fixing and cover pricing shows that the UN Global Compact has a role even in the UK. Unfortunately, like a lot of these things, there is no teeth behind the Global Compact.

The Equator Principles

This is, to quote the official blurb, 'a benchmark for the financial industry to manage social and environmental issues in project financing'.[2] In essence, the Equator Principles suggest that if you are investing in a country without effective national social and environmental legislation, then use the principles as a guide and do not go for exploitation and other unprincipled activities. The difference to the UN Global Compact is that the Equator Principles are aimed at project financing by banks and other financial institutions. Banks and other bodies that adopt these principles should therefore not invest in projects over US$10 million unless these conform to the ten principles. My (hopefully) jargon-free interpretation of the principles look something like this:

1. **Review and categorisation** (yes, these people are also bureaucrats): this is about the banks reviewing the environmental and social due diligence of the finance with emphasis on the impacts of the project.

2. **Social and environmental assessment**: this is for the borrower to carry out to the satisfaction of the lender.

3. **Applicable social and environmental standards**: this is basically about applying specific guidelines in countries that are not in the

OECD.[5] The guidelines are developed by the World Bank (parent of the International Finance Corporation).

4. **Action plan and management system**: the borrower has to prepare a plan to mitigate impacts and risks identified in the assessment. Again, this is the approach for non-OECD[6] countries, otherwise, national laws apply.

5. **Consultation and disclosure**: this is to make sure that the people impacted by investment are consulted in 'an appropriate manner ... to ensure free, prior and informed consultation'. The consultation will also need to be made public.

6. **Grievance mechanism**: this is to allow the people impacted by the investment to complain about it and have their concerns addressed by the borrower.

7. **Independent review**: this is to allow independent review of the assessment (number 2 above); the action plan (number 4 above); and the consultation process (number 5 above).

8. **Covenants**: this covers regulatory compliance of the host country, sticking to the Action Plan (number 4 above), providing reports and carry out any decommissioning as agreed.

9. **Independent monitoring and reporting**: this is where the borrowers pay external specialists to verify what it has done with the investment and report back to the financiers.

10. **Equator principles financial institutions reporting**: this is where the financiers reports publicly on their equator principles investments.

The Equator Principles can be seen as taking on some of the roles of social and environmental legislation commonly encountered in OECD/rich countries. However, putting the onus on the investor is a good idea to ensure some

5 OECD is the Organisation for Economic Cooperation and Development – in other words, the rich countries. Find out more at (www.oecd.org).

6 Apparently, this also includes OECD countries that are not designated as 'High Income'– told you they are bureaucrats.

compliance. The relevance for construction in the UK is perhaps less evident, but if you borrow money from banks that have signed up to the Equator Principles, then you may need to comply. (Of course, in a credit crunch, not many people are able to borrow or lend money, despite UK banks being the original group that set up the Equator Principles.)

At a national level, the rich countries have legal frameworks that dignify the individual and compel corporate behaviour. Although not many go as far as the Himalayan country of Bhutan, where the 'Gross National Happiness' is part of the national plan. The UK government has its sustainability principles which include 'ensuring a strong, healthy and just society' to describe the social elements of sustainable development.[7] This principle is about 'meeting the diverse needs of all people in existing and future communities, promoting personal well-being, social cohesion and inclusion and creating equal opportunity for all'.[8]

Social Sustainability at the Ground Level

For all the lofty top-down comments, what is social sustainability as seen from a construction project? Clearly, it is not likely that many clients and contractors will adopt the Gross National Happiness as an index of corporate performance (although individually, most will agree that happiness does matter). Nevertheless, much of the Global Compact and the Equator Principles are informally part of the way we do business in the UK. The types of activities that can be catalogued into this rather wide topic can include, for example:

- Stopping construction workers wolf whistling at women passing the construction site;
- Providing more than the minimum in site welfare facilities;
- Implementing an adequately budgeted health and safety initiative;
- Employing local people whether unskilled or inexperienced;

7 This is part of the UK's Five Shared Principles of Sustainable Development, whose emotional impact I shall explore in Chapter 8.

8 Visit the UK government sustainable development website (www.defra.gov.uk/sustainable/government/) for more.

- Buy materials locally;
- Promoting healthier life style through an information programme on site;
- Facilities for people coming to work on bicycles with secured bike sheds and so on;
- Offering paternity leave or flexible working for office staff;
- Supporting the local community through Section 106 funded activities;[9]
- Offering staff time off to support local community actions;
- A Corporate Social Responsibility (CSR) or Community Investment strategy that enshrines social sustainability activities as part of the corporate mission;
- Design a development or a regeneration programme that takes into account the social needs of people living in the area; and
- … (you can extend this list with whatever you deem necessary for your particular project).

Reality Check for Social Sustainability

There is one common factor in the list above besides social sustainability – they involve some form of behaviour change from 'business as before' to 'business for sustainability'. Change, other than stopping the wolf whistling, tends to cost money, and the list of activities above can add up to a significant cost. Whether your company or client is happy to pay this because they believe it is the right thing to do or treat the costs as a sustainability levy, money still needs to be spent. I will explore reducing these costs in Chapter 8 and behaviour changes in Chapter 16.

9 Section 106 Agreements are from the Town and Country Planning Act 1990. This allows a local planning authority to encourage developers to, well, basically spend some money on doing good in the community as part of their development project.

Although many projects carry out the activities listed above, the scope of social sustainability can be confusing for people working in construction as well as people from outside the industry. There are procurement issues (local buying); social issues (local employment); social engineering issues (eat more healthily); gender issues (wolf whistling); grand strategic issues (coordination with local regeneration); and marketing issues (greenwash).

Where do we start in a discussion on social sustainability? Do we start with the grand gestures – where a company expresses its deep and sincere commitment to the local society whether it is practical or not? Or do we start with the pragmatic actions – where a company graciously carries out activities that society is pressuring it to do anyway? Do we start on site? At the office or somewhere else?

A simple way to assess these needs is to use the criticality analysis described in Chapter 1. An adaptation for social sustainability is shown on Figure 4.1. The original Criticality Analysis for Sustainability (see Introduction) has three elements for the social areas:

1. **Committed parties**: the employees and investors. These people either committed their money or their livelihoods with the company.

2. **Involved parties**: suppliers and customers. These people essentially have some form of contractual relationship with the company.

Both these groups either pay the company or get paid by the company (or they deliver value to the company or get value from the company). Whereas the third group has no contractual relationship with the company at all. These people are the:

3. **Interested parties**: the broader society and its regulatory bodies. These people have no real links with the company but exert an influence which may be regulatory, social or emotional.

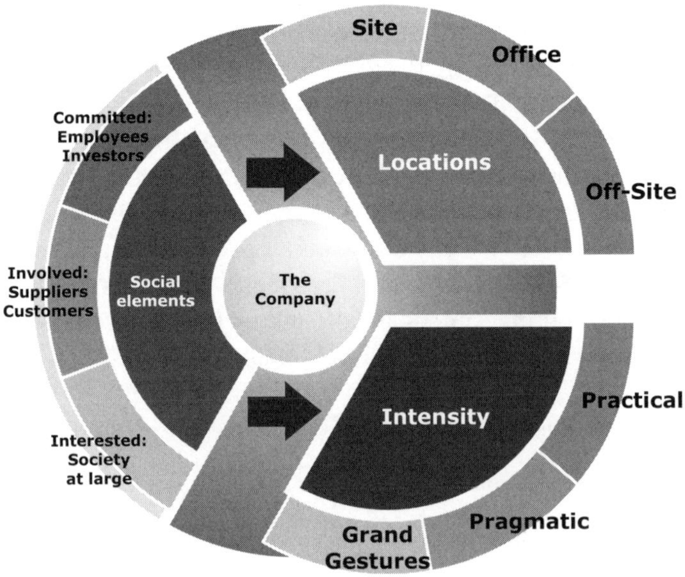

Figure 4.1 Criticality analysis for social sustainability

(Trying to) Keep Everyone Happy – Criticality for Social Sustainability

How are social sustainability actions going to be applied to these three groups, bearing in mind that we also have to do some constructing during a project? How do we use our sustainability actions and add value to them to make a real difference? Lovelace suggested that we should start with the premise that 'happiness is the foundation upon which we build our needs'. In other words, if we are going to make these three groups happy, what do we need to do? By considering 'increasing happiness' as a possible social sustainability mission statement. What does that mean for the committed, involved and interested groups?

Committed to Happiness

The committed group comprises investors and employees, it is a no-brainer that the basic ingredient for happiness will be a good return on investment and continued employment respectively.

A reality check tells us that it is unlikely most investors in construction companies are doing it for altruistic reasons – if you do, you might as well give money to charities directly. So investors want to see a decent return on their investment. Decent is a relative term – if another sector offers better returns then some investment will move to those sectors. Remember those construction companies that reshaped themselves into service companies to run PFI contracts recently? Many did so because 'service' attracted a better rate of return than 'straight' construction. Essentially, investors wants to get ever better returns on their investment and often they are prepared to pay an initial premium if they believe there is assurance of a bigger pay-off.

Employees seek a number of things as shown in Maslow's Hierarchy of Needs (see Figure 4.2).[10]

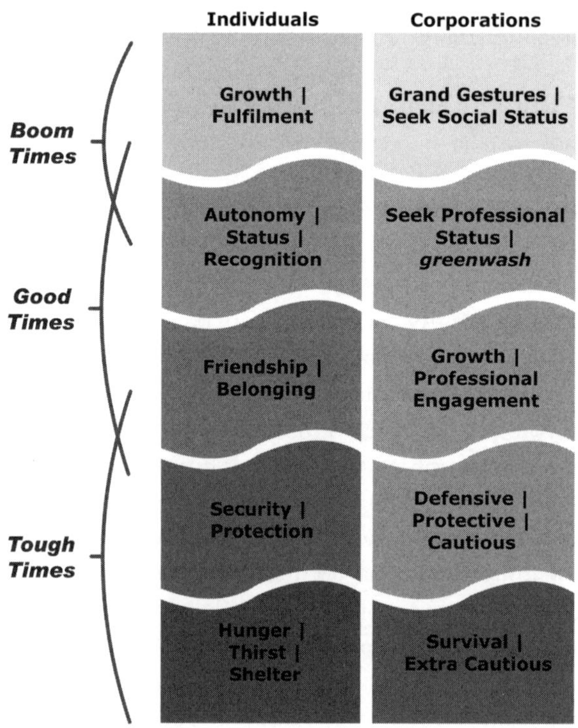

Figure 4.2 Individual and corporate focus

10 Abraham Maslow, a psychology professor, first reported this in 'A Theory of Human Motivation' published by *Psychological Review* in 1943.

During the good times, people tend to focus on self-growth as life is more stable and there is time to think about self-growth and development. But during the tougher times, the stability is gone and the focus switches to survival – hunkering down and sitting out the storm. People's needs shrink back to the basic ones of shelter and food – i.e. staying employed.

Involved in Happiness

The involved group are basically on contracts of one sort or another as either suppliers or contractors. A third group of involved relationship is a partnership that is neither supplier or contractor – much like how my sustainability project NoWaste[11] is working with ConstructionSkills at the big construction project in Stratford (which cannot be named because I am not a sponsor). The ingredients of happiness for these relationships will mostly be various flavours of 'hassle-free success': on-time; to specifications; and to cost.

For these two or three groups, an ability to just deliver on time, to specification and to cost together with avoiding embarrassments – in other words, just having a good day with no major or minor disasters, will probably contribute significantly to the happiness quotient.

Interested in Happiness

The interested group is perhaps less easy to define principally because it comprises everyone with their different needs and demands. This group can include society at large as well as the local neighbours of our projects. Nevertheless, happiness for this varied group can probably be described as 'no news is likely to mean good news'. In other words, if the construction industry can avoid the headlines: accidents on site; collapsing buildings; Office of Fair Trading inquires; or cowboy builder antics, the interested stakeholders are likely to find something else to be interested in.

All these different needs can be wrapped up as basic social sustainability actions: look after the people you deal with and being respectful to your neighbours. Just doing these well and you can be assured you are meeting a good proportion of their needs.

11 NoWaste is a Lean/sustainable operations programme which is focused on, as you may have guessed, reducing waste. (see Chapter 12 and the accompany CD)

Social sustainability therefore becomes a somewhat easier goal to aim for once we understand that it is actually quite simple. It, as Lovelace suggested, is the integration of the many things we are already doing: training; enhancing work–life balance; toolbox talks; recruitment; health and safety; welfare facilities; community activities; and so on. An added bonus is that each of these also contributes to building a better society and providing more opportunities for people in that society.

When you combine these activities to the other sustainability elements of financial and environment management, it becomes clear that just as finance is an enabler, social sustainability is also an enabler. Achieving the Triple Bottom Line, described in Chapter 1, will not really be possible without dedicated people and some budget. If people are happier, then everything just get achieved a little bit easier.

Since this book promotes the view that sustainability belongs to everyone, rather than just senior decision makers, a bottom-up approach is recommended. The social sustainability checklist I propose is the Lovelace List, an action-led approach described in Chapter 13.

The End-Users of the Sustainable Buildings

One aspect of social sustainability in construction is the treatment of the end-user. I do not intend to go into this deeply as my focus is in managing sustainability in construction. During the writing of this book, I was told by a sustainability specialist that, in his opinion, some architects pay lip service to sustainability. 'They are only interested in ticking boxes,' he said. He then gave me examples of automatic PIR sensors[12] on room lights but with no daylight overrides. Therefore people working (and walking) in an office in broad daylight will always have all the lights on. Even if these are energy saving lights, you are still wasting electricity. My colleague then screamed his frustrations about another one of these modern offices. There is one switch on the wall and it looks like a light switch. Therefore people flick it as they leave the room but instead, this switch turns off the PIR, so the light stays on. Then my colleague told me that there are still large new open plan offices with one thermostat for the whole floor. I thought the old Energy Efficiency Best Practice Programme[13]

12 Motion detectors to you and me.

13 A 1990s UK government initiative to save energy in businesses. This programme was a major supporter of technology solutions. I looked after the training portfolio of this programme

managed to see off this type of control system designs, but apparently not. It seems that no matter what the codes specify, my colleague reckoned there will always be designers who take the view of 'This is only a job and I will never work or live in that building.'

My colleague's cynicism and frustrations about some of our colleagues were confirmed by Mindy Hadi of BRE, a consultancy. Hadi told me that 'a building is a system and that includes the people occupying it'. She further adds designers 'cannot design a building as though it is an empty box in isolation'. She believed that 'if a building does not work for its occupants, it is not sustainable no matter how green it is'.[14]

The Recession and All That – Caveats for Social Sustainability

Part of this section was written before the construction industry fell off a cliff in mid-2008. Since then, thousands of jobs have gone and many are still going despite the reports in the trade press about developers starting to rebuild their land banks and house prices creeping up in the UK from summer 2009. In the summer of 2010, the national news reported that the UK economy had grown at a much higher rate, thanks to the construction industry. While business is booming, it is easier to take on additional non-core activities as part of the business activities. By this I mean anything that does not add to the functionality of the building, such as the social elements in sustainability. When times are tough, whether it is recession or the early stages of recovery, is there sufficient justification to cut back on everything that is not part of the core activities?

This is a tough one to answer. If everyone is doing it, then it becomes a stampede to the lowest common denominator. The situation then becomes a matter of comparative advantage – the socially aware company will have an advantage of its reputation, but also suffers from higher costs. It is almost a matter of who can hold their breath longest while being under the flood of recession. Of course, this is the case in a 'perfect' competitive environment, but not necessarily the reality in the UK. Our market situation can often be distorted with local authority planning obligations. The use of the Section 106 Agreements

between 1992–2000.

14 BRE even has a publication showing some of these less-than-good practices: 'The Move to Low Carbon Designs: Are designers taking the needs of building users into account?' M. Hadi and C. Halfhide, FB21, *BRE Trust*, 2009.

also brings added social sustainability aspirations (or complications) into the process. So we muddle through, recession or not.[15]

Besides the headline items on the social sustainability agenda, there is a host of related issues that are not always attached to sustainability but have strong implications for a full transformation to sustainable construction. There are two strands to sustainable construction: build sustainably; and build sustainable buildings. The former is really the way forward but this requires changes in behaviour. The latter is just meeting a client desire and it is how a lot of companies treat sustainability. However, whether you are a designer or a builder, there are opportunities not just to build sustainable buildings, but also to run your business in a sustainable way – the office; the project site; and dealing with people (the committed, involved and interested).

Social sustainability is not just about having checklists, 'we build to level 4 of the Code for Sustainable Homes' or 'we have a BREEAM certificate', it is also about the relationship with people during the process of building. Making people happier should perhaps come with a caveat: make people happier so that they can become more innovative. People in the industry needs to be motivated so that they think a little beyond meeting the letter of the current requirements.

Rebecca Lovelace finished her discussion with me by suggesting actions to take us forward towards integrating social sustainability more strategically into our industry:

- Carry out an audit of all community investment activities taking place within the company and on construction projects. It's amazing how often much of what is happening is not captured or formally coordinated.
- Think strategically – is there capacity to do more? Which activities would bring greater benefit to both our company and their communities?
- Leverage your combined expertise by working in partnership with communities.

15 I say this because only two of my sustainability manager acquaintances were made redundant during the 2008–09 recession. Maybe most of my friends were lucky, or perhaps their employers really believe in their own PR!

- Develop and implement a strategic and integrated community investment strategy.
- Ensure that there is adequate public measurement, evaluation and reporting on all community investment activities.

She recommends that we ask ourselves four critical questions:

1. Do I respect the people working with me?
2. Do I recognise their contributions?
3. Do I reward them in an open and fair way?

And perhaps most importantly:

4. Do I want to make people happier?

A 'yes' answer to any of these questions gets you started on social sustainability. These questions also form the basis of the 3R tool: respect; recognition; and reward. This is described in the Chapter 16.

In Chapter 13, I will examine how the various ideas from this chapter are turned into a useful checklist for the things you can do to make social sustainability happen easier under an adverse (or improving) economic climate.

5

Capabilities, Skills and 'Green Skills'

A major concern in the UK construction industry is skills shortage. This has been blamed for everything from poor workmanship, migrant workers and the high accident rate. Not withstanding the continuing efforts of ConstructionSkills, the Sector Skills Council, in trying to attract more people into construction, it seems the debate will continue. One interesting thing about this is that although the debate goes on year after year, buildings are still going up in the UK (although the recession has slowed this considerably).

A Skills Shortage in Construction?

The media, both general and the trade press, is constantly reporting on the skills situation in construction and in the country generally. Depending on which bit you read and also when you read it, it may be true or it may not be true. Is it real?[1]

Generally, the demand and supply of an economy is self-regulating. If there is a lot of money to be made in financial services, people flock to it. If there is demand in construction, then people will similarly flock to the industry. However, in the UK, this does not appear to be the case. Both *Building Magazine* and *Construction News* regularly report on the 'skills shortage' and what ConstructionSkills is doing about it. Every so often, company bosses are quoted as saying that they cannot bid for work because of a shortage of skilled staff. The CIOB, a professional body for the industry, publishes reports annually on various issues in the industry and the concerns of a lack of skilled

1 This really depends on which week's trade press you read. I followed it in some detail over 24 months for this book and am sort of convinced that yes, it is true this week. But then let's wait a week, maybe not.

personnel is a regular item on the list. Now why does our industry not respond in a regular way towards supply and demand? Is our industry really so unique that it can defy one of the basic principles of economics?

I think there is a bit more to the construction skills shortage. Let us examine how shortages start. Usually it is a case of supply not meeting demand. So why is UK society ignoring our industry? It may be because of the low rewards in our industry, or perhaps a lack of job security or some other factor, such as that having proper construction skills are not the same as an ability to DIY. Professional skills take a while to learn and perfect.

The rewards issue is both the easiest and hardest to tackle. Easy because if you pay more, people will come. Hard because if you pay more, then profits go down.[2] Let us try again. If you pay a bit more, perhaps a few more people may come. Easy? No, because why pay a bit more when there are migrant workers who will work for less – in fact, on minimum wages? The alternative to paying migrant workers low wages is paying a 'decent living wage', and constructions companies probably do not mind provided that they can pass on the costs to the client. However, many construction clients, other than those on national prestige projects, tend to be somewhat frugal about this type of costs.

Even that may not solve the skills shortage immediately because it will take a while for unskilled people to be trained and then join the workforce. If the timing is wrong then the newly skilled join the industry in a downward part of the cycle and they all get laid off.

This gets to the next point on the skills shortage list, that of job security. Since construction is cyclical, then the industry will have boom and bust periods. In boom times, there are never enough skilled workers, during the bust periods, there are not enough jobs to go round so people stop recruiting and training since there is no call for them. The current boom–bust cycle has just entered the bust stage from a boom that can be tracked back to the mid-1990s. I do not have trade press reports of skill shortages from that period, but some sources appeared to demonstrate a 'non-shortage' coinciding with the recession in the early 1990s.[3]

2 In other industries, an inability to satisfy demand due to skills shortage may cause problems, but the subcontracting culture in UK construction means there is a lot of experience in outsourcing for skills. This is shown by the fact that despite the always-present skills shortage, projects are still being completed.

3 Ralph Morton's book, *Construction UK: Introduction to the Industry* is a useful guide full of interesting information. It was published in 2002.

There are many instances of skilled professionals knowing they are in demand and behaving accordingly. In specialist activities and at particular geographic locations, some operatives and craftsmen do behave like *prima donnas*: 'Why should I wear PPE? I can get a job tomorrow down the road without these #@&* rules.' We have all met people in the industry like this and that is often when we moan about a skills shortage because otherwise we would not have to put up with idiots like him.

Even before the recession, despite the encouragement in the media about how much a plumber can earn, construction skills are not readily learnt in a short time, as many who engage cowboy renovators found out. A difficult to learn skill coupled with both a lengthy period of learning and a short-term industrial outlook all contribute to construction companies declining to invest in training and development. After all, why train when you can hire skilled people ready to work immediately? Even better, why pay skilled UK nationals a 'fair' wage when you can hire skilled workers from Europe's own Third World for minimum wage? Despite Sir John Egan's correct assertion that 'construction does not yet recognise that its people are its greatest asset and treat them as such',[4] a common view seemed to be 'so what?' followed by some complaint like 'it's ConstructionSkills' fault'. Even when training is being delivered to the unskilled, there are still complaints: in my NoWaste sustainable construction project at the big project in Stratford, there were complaints about unskilled people trained by BeOnsite, Lend Lease's training arm, leap-frogging others with years of experience.[5]

So what if we cannot get enough UK people into training? We have skilled professionals from Eastern Europe and overseas to fill the gap. In a strictly economic analysis, there is nothing wrong with this. The other countries supply the UK with what they are best at doing, in this case, lower-cost skilled workers and the UK gains the benefit of lower costs and buildings are still being put up. In fact, having migrant workers makes it easier to lose people during the 'bust' part of the business cycle – the migrants pack up and look for work elsewhere.

In terms of sustainability, the numbers seem to stack up too at first: economically it makes sense; socially the migrant workers earn more in the UK than in their own countries. Environmentally, there are concerns about 'foreign workers' not understanding our rules about pollution and legislation, but the

4 See Chapter 18 for more on Sir John Egan's Rethinking Construction Report.

5 'it took those people with BeOnsite 10 weeks to get a plant operators ticket, it took me years to work my way up'.

industry does have multilingual signs and guidance. My own experience of overseas workers showed that while they may not be able to speak English fluently, they are often more aware of the need to cut waste.

The other side of the sustainability coin is a bit tarnished. When migrant workers are being paid less than their UK counterparts, it is unethical and wrong. It is essentially saying the migrant workers do not deserve UK wages. The problem it seems, is that most people in the country prefer a fatter wallet. So we condone the lower wage for migrant worker since it means our buildings are a bit cheaper. (Of course, paying more will start one of those tensions: can we afford to pay our people more when the lowest cost bid is often the winning one?)

Whether the companies themselves are interested in developing the necessary skills is an entirely different matter. Although progressive companies are regularly wheeled out to declare that investment in skills does reap benefits, this does not alter the fact that a lot of companies do not really subscribe to it. They also do not want to pay for it. A few years ago I was working with the CoRE programme[6] in East Anglia with union partners to help construction teams to reduce waste. My union colleagues advised me to put a list of legislation in the brochure because that is the 'only way employers will let their people attend a training programme' (even when it was free). Indeed, many employers also complained about paying the travel cost of their teams coming to our free seminar to learn about working more efficiently. Similarly, my NoWaste programme at the big Stratford project had to require a lot of cajoling, reminding and hassling to get people released for sustainability training.

In other companies, the will to develop their teams is there but the actual task of catching up with the latest technologies and regulations can be quite daunting, as Andrew Bowles and Alan Shingler told me about their never-ending efforts in making sure all their colleagues at Sheppard Robson are kept at the cutting edge, recession or otherwise. Another colleague, a buyer, told of 'a certain learning curve towards a more sustainable way of construction'.

Pat Dowling, a UCATT convenor,[7] told me that he believed the average construction worker needs to have more information so that they know what

6 The Construction Resource Efficiency programme was managed by BRE, a consultancy, in England 2006–08.

7 UCATT is the national trades union for construction: Union of Construction, Allied Trades and Technicians.

is going on. Pat's view, which I fully support, is that much of the sustainability information is aimed at managers and specialist professionals. The construction operative does not get a lot of details and therefore they do not have a lot of information on the consequences of environmental damage, and the cost to ecology and society of construction activities.

For construction operatives to take their share of the responsibility for sustainability, they need to have the right information to help them make decisions. For example, Pat suggested that operatives should have the following information on materials they use: source and end of use disposal; potential for environmental damage; and also the cost of waste disposal. I always felt we need to respect our workforce, encourage innovation and more effective working. For people to do that, however, they need sufficient information so that they can make appropriate decisions on sustainability. This was part of the NoWaste sustainable construction training since 2004 and more recently, I introduced a new element: common knowledge in sustainability – a set of questions in the CSCS touch screen style on sustainability[8].

Another problem is that many companies are small, and small companies have a different approach to training. It is not that small companies do not believe in training, it is just that formal training is expensive. Many of the very small construction business I talked to are interested in taking on a 'young lad' and many have. Some of the craftsmen I engage to carry out refurbishment domestically nowadays are 'second generation' craftsmen, trained by their mentors who first worked for me in the 1980s and 1990s. On the job training does exist for the very small company and they will pay for training that leads to compliance and accreditation in a critical job area. The willingness, however, often stops at the type of training that is not an accreditation requirement and also costs money.

At the highest level, things can also be disappointing too. I was invited to run a session at a sustainability leadership course at an elite UK business school in 2007. The delegates at the inaugural programme included people from all walks of industry: hotels; dog food; fire brigade; charities; and the construction sector. All are hard-nosed and pragmatic managers who wanted something to take back on Monday to show their colleagues what the expensive course did for them. It was therefore a bit of a surprise when the course turned out

8 See Appendix 2 and the accompanying CD for the full Common Knowledge in Sustainability (CKiS) questions. At the time of writing (autumn 2010), the Environmental Agency was also preparing its own set of questions.

to have a big dose of navel gazing, taking off the shoes and hugging. There was a lot of talking about self-discovery as well rather than brainstorming for managerial approaches to deliver sustainability. My approach of tackling day-to-day issues, setting up a strategy and so on, did not fit in with the other trainers' journey of self-discovery at all. It is disappointing that a top business school can be so out of touch with the daily concerns of these senior managers – the entire programme felt like a late 1980s team bonding exercise. There is a lost opportunity here somewhere.

This is an issue that also concerned Phil Henry of Polypipe, a manufacturer of water management solutions. Henry believed that the skills shortage is not just the need for the right skill set, but also the right attitude as well. Henry's view was that professionals need a bigger picture view and not just fulfilling the technical aspects of a job.

Echoing my colleagues, I also do not think we really have a skills shortage problem but more a vacuum in thinking about the strategic benefits of training. The skills shortage is a result created by the industry itself. And as long as construction companies leave it to others like ConstructionSkills to sort it out, the skills issue will remain.

Whatever the cause, there is a serious issue about training people in construction, one which hiring migrant workers is not likely to be able to help much. This is the skills needed for sustainable construction. It is new, vitally important and just about everyone in construction will need to know it. Only problem is, well, we do not really know what sustainable construction skills are.

A few years ago, at a recent meeting with a government agency, the host mentioned the need for 'green skills', and everyone at the table (including me) nodded agreement and made appropriate murmurs about its importance. Afterwards, I thought about it and realised I have no idea what green skills are and what should be part of a green skills package. Yet we all talked knowingly about it as though it was very clear. I have since been thinking and testing my ideas on skills, capabilities and these started with the following question.

Just What Are 'Green Skills'?

A traditional description of 'green skills' can include a whole range of knowledge and competence in environmental issues broadly categorised into four main areas (see Figure 5.1).

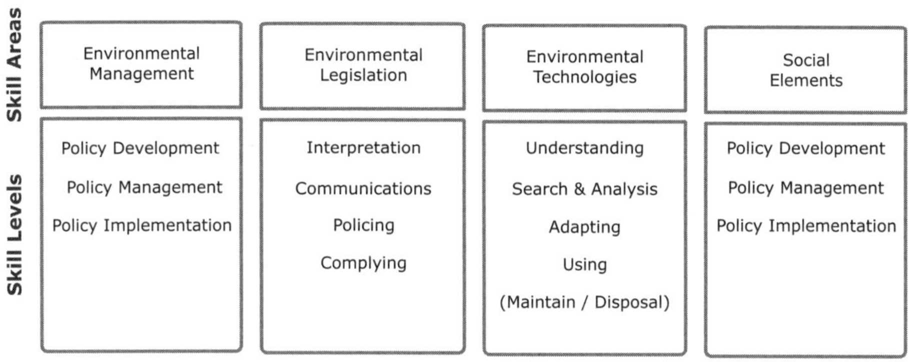

Figure 5.1 Traditional Green Skills Matrix

Environmental management: this is the planning and managing day-to-day activities to enable an organisation to deliver environmental benchmarks and targets.

Environmental legislation: this area covers the understanding of the various laws and what they mean for the particular company. Once the interpretation is done, many of the follow-on activities is made up of telling people about the laws and making sure people stay within the law as they work.

Environmental technologies: this is about the various energy and environmental 'kit': biomass burners; wind generators; motion sensing lighting controls; and super cements that absorb carbon dioxide; and so on. Since new technologies are constantly being developed, new products come on the market regularly and even the specialist can find it difficult to remain completely up to date.

'Social' elements: this area of skills is about planning and managing actions to support the environmental management activities.

Most of these skills are critical and important for specialists – they need these to carry out their daily work. The rest of us, however, only have a supporting role to play in these areas. Being told what to do is not really 'okay' in an area where we too want some degree of ownership.

Not everyone needs all of these skills to do their job. In many instances, the policy developing skills, for example, are not required as most people will only require 'doing' skills. Similarly, besides the legal specialists, most project teams will probably need a working understanding of the legislation and only with the elements related to their work.

The skills for managing environmental technologies rely on an understanding of the context of the technology. The specialists will need to find and assess the most suitable technology, and also how to adapt it for deployment. The users need to understand how the particular technology can deliver the desired results (or knowing which button to push). Finally, there needs to be an understanding about the disposal at the end of the product's useful life (or know the phone number of your waste company).

Nevertheless, the traditional set of green skills needs rethinking because its scope is too limited and ultimately, it may not be sufficient to help the rest of us 'do sustainability'. Although many of these green skills areas are currently being practised very effectively by construction sector professionals; architects and environmental managers do not actually staple gun the plasterboard or pour the concrete. For the construction operative whose actions really define sustainable construction, there appears little targeted training other than some of UCATT's[9] sustainability programmes and my friend Andrew Pear's Cafe Van programmes.[10]

Pat Dowling of UCATT reminded me, 'My members want to contribute, but they do not always know what to do. They do not have any knowledge in this area.' To counter this, a sustainability manager from a major construction company told me, 'we just tell 'em what to do, they're good at being told'. This situation is not ideal, but is possibly workable if the sustainability expert knows exactly what the job involves and can provide precise guidance on every step of the process, including unexpected events. However, this is unlikely and

9 UCATT's Project Build and Project Sus'd, are good examples of sustainable construction training programme that I developed and aimed at team leaders and operatives.

10 The Cafe Van is a visiting sustainable construction training service developed by my friend Andrew Pears of Kotuku. It provides the much needed practical training for managers, supervisors and operatives on 'doing sustainability'. (also see Chapter 12)

furthermore, denying the workforce the opportunity to take the initiative also means that you are not going to get innovations from the people actually doing the work. Nevertheless, being told what to do is easy for both sides, the 'telling' side passes on the responsibility and the 'listening' side forgets as soon as the boss' back is turned.

Not 'Green Skills', But Capabilities

At the time of writing (autumn 2010) a group of London Universities and Colleges are developing a new set of skills for sustainability. But I believe we should not develop more sets of green skills but rather focus on about the professional capability to make sustainability happen. We need everyone in the industry to have the practical ability to do sustainability. The shortage in our industry is not necessary skills or green skills, but some very basic professional capabilities. Just what do we really need? I think at the minimum, everyone should be able to:

- reduce waste
- minimise environmental impacts
- comply with the legislation.

The three areas described above are three essential capabilities, not skills. For example, to make sustainability happen on site, the teams will need to know not just how the waste is managed, but why it occurred in the first place and how to deal with the waste such that it does not happen again. This is the approach taken by my NoWaste programme and the focus on the relevant context for the people involved.

Instead of legislation or environmental management, it is waste that is at the heart of sustainability. Why? Because waste means we are getting it right only some of the time, not getting it right all of the time. Waste means extra resources being required to mop up the mess and sort out the defects. Extra resources mean extra burden on the environment. In construction, just about all the raw materials besides timber and plastics comes from roasting, baking or melting rock of one sort or another and that takes a lot of energy.[11]

11 In a very simplistic way: cement is made by roasting rocks; brick is from baking minerals; glass and mineral fibre insulation are from melting rocks.

A part of construction is not called the biblical trades[12] for nothing, they have been around a very long time. Nevertheless, there are new ways of doing things to minimise environmental impacts and construction activities ranging from design to the 'Method Statements' on site need to reflect this.

Compliance with legislation is increasingly more important not just because of the environmental impacts but because of the financial impacts. Being fined for environmental infraction means spending money where you really do not need to (or even want to) and small environmental mistakes, such as sending out some unsegregated waste in a skip for landfill, can also mean a hefty handling fee as well as the Landfill Tax bill.

Each of the three capabilities is made up of: technical knowledge; experience; problem solving; and initiative.

We need the technical knowledge to allow us to assess a situation at work. Our experiences help us to classify the information and our problem-solving knowledge helps us deliver solutions. And most importantly, we also need to have a workplace environmental that encourages us to take the initiative rather than just walk by and ignore a problem.

Letting the workforce to take initiative is something that companies both desire and fear. It is desirable because we all believe (or have been taught to believe) that innovations lead to a continued competitive advantage. It is feared because we have also been taught that 'managers do the thinking' and if everyone is coming up with ideas to make things happen, then managers will lose their power. The concern here is to make sure 'workers must not get control'.[13]

The top companies know that initiative is desirable and most importantly, they realise that, if they cannot motivate and mobilise their workforce to take the lead in bringing about sustainability, then it will not happen at all.

These three capabilities are the starting point for everyone working in construction to be able to 'do sustainability'. The knowledge requirement

12 Traditional trades like joinery (carpentry), bricklaying painting and scaffolding are known as the biblical trades to distinguish them from heating ventilation and air conditioning (HVAC) or lift engineering.

13 A senior manager of a trade confederation actually said this to me several years ago 'we must not let the workers get control of sustainability, letting them run health and safety was bad enough'.

for this does not mean becoming a rocket scientist, but on how to observe and how to solve problems. These allow everyone to make a contribution to sustainability.

Having explored what is needed to make sustainability happen, then the logical question is that whether it can happen in our industry.

Can Construction Really be Sustainable or is it Just Marketing Hype?

There is more than one answer to this question because there is more than one construction industry. There are the high profile commercial and government projects, then there is house building and then there is refurbishment, which can be further split into commercial refurbishment and domestic refurbishment. The answer from each of these sectors will vary but is likely to contain some of the following sentiments:

yes, but some builders need to be trained first ...

yes, but the client will have to request it ...

yes, but the supply chain needs to be up for it ...

These all contains a somewhat unsubtle 'code', namely: 'who is paying for it?'

Training costs money and the industry structure of self-employment and subcontracting does not always encourage lifelong development of skills in the larger contractors or even for regular upskilling in the very small operations.

Clients in commercial and government projects are already asking for sustainability as part of the construction package. Since sustainability is essentially a requirement for doing business at this level now, then the costs are part of the accepted cost of doing business. This is generally the case for top-tier, high prestige work. But work like this is not representative of the entire industry.

For example, in domestic refurbishment, the situation is less clear. Some householders are progressive and willing to pay, whereas others are more cost constrained and may opt for a less sustainable solution. There is a further issue in domestic work in that many people are reliant on their contractors to recommend the relevant technologies. However, many small contractors are not necessarily up to date with the latest technologies nor are they always competent in installing them. This may then result in the contractor recommending the tried and tested traditional technologies.[14]

The supply chain has been a good place to pass the buck. Many major contractors are implementing procedures to reduce the number of suppliers with the aim of raising the quality of the subcontractors through the pre-qualification requirements. But the system of multi-tiered subcontracting can mean the actual person doing the work is no better informed than the domestic contractor (and may even be the domestic contractor). This is especially true for those trades that are in demand and have insufficient number of qualified or accredited operatives.

Can construction be made more sustainable? In the absolute sense that every commercial activity has an impact on the global environment, then nothing is going to be sustainable. Since building anything requires energy and natural resources then no matter how many wind generators we build, the original impacts will remain. However, we can do better. We cannot go on generating 50 per cent of the UK's CO_2 emissions or creating 100 M tonnes of waste each year.[15]

Nevertheless, people need buildings and we are there to build them. The important point is to do what we do with the minimum waste. By being efficient, we can minimise the damage to the global environment. Construction can be made more sustainable, but it is unlikely to be completely sustainable with the current level of political will, public support, technology, investment and managerial capability. Of course, we can always make ourselves feel better by planting several hectares of forest in some foreign land to soak up the equivalent CO_2, but as consultant Deltcho Vitchev of Renaissance Finance International said, 'it is not far off the medieval practice of buying indulgences'.[16] Offsetting is good for the soul (and PR), but it does not do a lot of good for the hole in the ground where the original limestone or granite comes from or where we dump the construction wastes.

14 My plumber and electrician both advise me regularly, 'you don't want to mess with those'.
15 From the UK government's Draft Strategy on Sustainable Construction, 2007.
16 Also discussed by George Monbiot in the *Guardian* newspaper, 18 October 2006.

There are several critical issues on skills and capabilities: wastes; capability; and money. Let us examine these in order.

Of the three, waste seems easiest to deal with, the government has plenty of legislation about it and every organisation has some sort of plan to deal with the waste. To many companies, it is about sending the waste off to be recycled and perhaps making sure that zero-net waste principles[17] are addressed (see Figure 5.2). But even recycling takes energy, not only the energy need to transform your product but also the energy to transport your waste and shipping the recycled product to the next user. Do not get me wrong, recycling is not a bad thing, but it is not a complete solution either.[18]

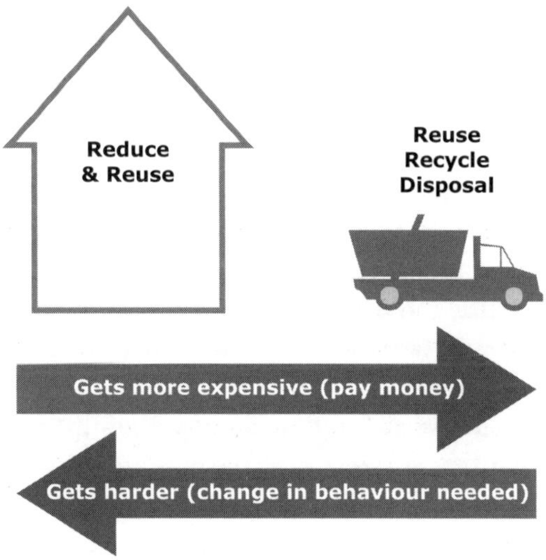

Figure 5.2 Managing waste: the options

I will revisit waste in greater detail in Chapter 8.

17 This is one of the UK government's strategic initiatives to cut waste by requiring that the value of materials for recycling and reuse matches the value going to landfill. It was in the draft Sustainable Construction strategy of 2007, but disappeared in the full 2008 version.

18 All the emphasis on recycling makes me wonder why the policy makers rarely talk about reduction – maybe they have no idea how it's done. But of course, reduced landfill will mean a reduction in Landfill Tax too, and the companies operating the landfill sites also have an interest in recycling – enlightened self-interest perhaps?

Capabilities can be developed but that takes time, just like planting forests for carbon off-setting takes time to grow. If we are not prepared to invest in this, then not a lot will happen.

Money is of course the big problem. After all, when you have money, problems tend to become a bit easier to resolve, and when you do not have money, every problem gets tougher. However, in a recession and recovery coupled with a credit crunch, where is the money coming from? Without money, can we move sustainability forward at all?

Skewed, Skewered or Sideways

Without money, sustainability can seem like a forlorn hope, especially as it is skewed towards the environmental elements as described by Guy Hazlehurst (see Chapter 2). The environmental-focused approach to sustainability tends to result in environmental management becoming a cost centre, not a particularly comfortable prospect during bad times when companies are pinching pennies. However, while the requirements for legislative compliance mean the environmental managers will not lose their jobs, some companies may not wish to go much more beyond mere compliance. The picture about sustainability in construction does not necessarily look healthy. Although no one will say this publicly, when companies think about reducing cost, activities that do not make money are not safe, with the exception of basic legal compliance.

How do sustainability professionals ensure that their authority and influence are not diminished with their budgets? This is a serious issue as it involves the future of our industry.

This is where *sideways* in Chapter 3 comes in.

By taking a step sideways from the over-emphasised environmental elements is a way forward to deal with the issues of efficiency, capability and money.

A move to a more balanced position is critical not only because it allocates a more appropriate amount of attention to the social and financial elements, but also engages the financial professionals who manage the money in the company. But most important of all, it takes sustainability out of the ghetto of environmental management rhetoric and brings it into the entire company.

Sustainability belongs to everyone, not just to the bosses, the specialists or the champions. Unlike previous management initiatives, everyone has some idea of what sustainability is. The benefit from this is that it is easier to get the corporate message across, but this also means that the ownership of sustainability is shared. However, many companies still take the view that technical knowledge and authority shall always remain with the managers and specialists. But more and more technology is being diffused to serve the needs of everyone else rather than remaining as a source of power for the managers and specialists. As for authority, if you always tell people what to do, then they will never exceed your expectations nor will they be particularly innovative.

By taking a sideways step, sustainability moves from being the domain of a small number of sustainability managers to something everyone can participate in.

The capabilities described above are already within the awareness of most professionals working in construction, whether managers, supervisors, craft workers or operatives. By helping people to use their knowledge and awareness in the right context, we unleash the potential for delivering solutions to make sustainability happen.

This is a major change when compared with earlier sustainability initiatives. The energy efficiency programmes were dominated by energy technologies where the technologists ruled supreme. Environmental management was confused with 'tree hugging' and by the time people start to be aware of it, early legislation created an 'elite' of policy makers and technical specialists. This time, thanks to the often breathless media reporting, a lot more people will know of the threats to the eastern parts of the UK by flooding because all the ice in the world will have melted into East Anglia. Furthermore, all the polar bears will die because tropical insects will migrate north, mutate into monster insects that eat polar bears, and of course, everyone will find it hotter. Sensational (and probably untrue and unlikely) as all this may be, these images contributes to a distorted public vision of why sustainability is important!

Now it is up to the sustainability specialists to engage and listen to their colleagues: what is doable, what is possible and what will never be accepted. They will also need to do some marketing, to make their ideas more acceptable, to gain support and buy-in for their programmes. They need to be coaches and facilitators for their colleagues. Most of all they need to create a suitable context so that their fellow workers can contribute and share ownership of the sustainability activities. This is a different way of doing things, not what we are used to. This is a different world.

6

Sustainability: Confusion in the Workplace

Having the right skills or capabilities is a good thing, but being able to deploy them effectively is even better. The key to effective deployment is about aligning the right components in a company in line with the corporate mission. To do this well, we need to make sure the corporate communications make sense to the people doing the work. In most companies, this is hard enough, but in construction, our subcontracting culture makes this even harder. Trying to align people from different companies, many of whom who are engaged with your company only for a short period of time, towards your corporate targets can be extremely difficult. In managing sustainability, it only takes one operative to do something wrong and you can have both the media and society breathing down your neck.

What is the corporate sustainability message supposed to do then, in both the good times and the less good times? I guess it is supposed to motivate, energise and keep up the spirits, besides boosting the company's standings with investors and society. It may also reflect the belief of the people who make up the message.[1]

Since we are working with colleagues who may have a different view on sustainability; as well as under pressure to do our day jobs and also to cut costs, the sustainability message had better be effective.

Many organisations deploy some form of corporate sustainability statements and slogans. Some of these are short and snappy (they probably hired expensive marketing consultants), some are short and incomprehensible

1 It is not easy for a company to produce a sustainability statement that reflects the views of its entire workforce or investors especially with the different views of what sustainability should be.

(they probably did not) while others are long winded and dull. All of them are worthy, of course, because they express a vision for a better world where the organisation can balance the needs of people, planet and profit.

Unfortunately, a lot of people view these visions as fantasy. Not because the visions are flawed, but perhaps because the message itself is inadequate for the task. In a work context, what is equally important to motivating the staff is the ability for the staff to carry out activities that delivers the organisation's objectives. Fuzzy messages do not help the staff in making day-to-day decisions on 'doing' sustainability.

I have selected two examples of 'corporate sustainability statements' to show how they inspire and enable us to 'do' sustainability.

A few years ago, the London Government[2] chose the following snappy (and highly polished) themes and called it '4R':

- Results (getting results)
- Responsibility (taking responsibility)
- Resources (managing resources)
- Respect (developing respect).

This is good stuff and it makes a good sound bite. But say you are trying to use this inspiring 4R to help you make a decision in your job – will this help you chose what actions to take?

- getting Results – yes, we all have to get results (*ho hum*);
- taking Responsibility – well, yes we subscribe to that, but it's really his/ her/ their fault, not mine – I'm just following orders (*yawn, ignore*);
- managing Resources – we always order more than enough in construction so that we do not run out because some consultant told us that if we suddenly run out of materials, getting emergency

2 'A sustainable development framework for London', London Sustainable Development Commission, 2003.

spares will cost us four times as much (*so we are doing sustainability – our way*);

- developing Respect – we do not get much from our management (*so why bother, ignore it*).

The 4R is a nice slogan and will look good in pamphlets, posters and bulletin boards, but it will not do too much as practical day-to-day guidance in making decisions on sustainability.

The UK government's 'shared' principles of sustainable development,[3] shown below, are worthy but quite dull.

- Living Within Environmental Limits – Respecting the limits of the planet's environment, resources and biodiversity – to improve our environment and ensure that the natural resources needed for life are unimpaired and remain so for future generations

- Ensuring a Strong, Healthy and Just Society – Meeting the diverse needs of all people in existing and future communities, promoting personal well-being, social cohesion and inclusion, and creating equal opportunity for all.

- Achieving a Sustainable Economy – Building a strong, stable and sustainable economy which provides prosperity and opportunities for all, and in which environmental and social costs fall on those who impose them (Polluter Pays), and efficient resource use is incentivised.

- Using Sound Science Responsibly – Ensuring policy is developed and implemented on the basis of strong scientific evidence, while taking into account scientific uncertainty (through the precautionary principle) as well as public altitudes and values.

3 You can find this and more (a lot more) at the UK government's website on sustainability: however, with the change in UK government in 2010, the DEFRA website is being updated as I write, so both policy or website may change. For what it is worth, this is where I found the five principles: (www.defra.gov.uk/sustainable/government/what/principles.htm) webpage accessed 6 March 2008.

- Promoting Good Governance – Actively promoting effective, participative systems of governance in all levels of society – engaging people's creativity, energy and diversity.

Despite being somewhat pedantic (references to other external elements, for goodness sake), the government, nevertheless, made it clear for a policy to be sustainable, all five elements must be respected – so at least they believe in a balanced approach. There are a lot of words here and I felt that this illustrates my point of dull but worthy. Like a lot of worthy documents, it can be embarrassing when events and conduct showed that the government did not really followed them: the press coverage of waste[4] by government departments shows that incentivising effective resource usage is not necessarily happening; the MP expenses scandal does not highlight effective systems of governance.

The question for me, however, is whether these words motivate or help me in reviewing my daily activities? Can I use them as a guide to how I work, deal with problems and 'do' sustainability? Unfortunately, for me, the answer is not really and the government's slogans are no better than London's 4R.

For example, how will I use 'living within our environmental limits' as a guideline to my daily work? For starters, I do not know what the 'environmental limit' in the things I do is, without that knowledge, there is little chance of doing much within some unknown limit.[5] This type of slogan is very good when projected onto a screen during a presentation, but more or less useless when you are making a judgement in practical situations. The 'sound science' slogan[6] is even more difficult to understand and apply.

4 In October 2010, Sir Philip Green reported that he believed that potential government savings can be in 'billions of pounds' as reported by the *Financial Times*, 11 October 2010.

5 Limits are always difficult to define and often overtaken by events. In 1798, Robert Malthus, an economist, predicted that if population were to grow at the then high rate, the world will run out of food. But that did not happen because the Industrial Revolution happened instead and mechanisation increased productivity. A more modern economist, Julian Simon, suggested in his book The Ultimate Resource that human ingenuity tend to save the day from doomsayers.

6 The two examples – the 4R and five principles are from the public sector, but the private sector has similarly inane slogans and boring prose. It makes you wonder whether the people designing these unhelpful messages really think through what they are for.

Table 6.1 Slogans and responses

	Examples	
Slogan	**Possible personal response**	**Possible corporate response**
Living within our environmental limits	What is an environmental limit?	We comply with legislation
Ensuring a strong and just society	I voted and paid my tax, isn't that enough?	Not our job – we paid our taxes!
Achieving a sustainable economy	A bit vague and large scale ...	Isn't that the government's job to set the policy?
Promoting good governance	For whom? The government does not set a good example.	We comply with the law ...
Using sound science responsibly	Huh? We are bombarded by all kinds of media science scare stories	Huh? Business always gets the blame

Why is all this so important? Can we not accept that the slogans are intended as political messages? Unfortunately, we already discussed how a sceptical and dissenting staff member can wreck your sustainability initiative, the last thing you want is to send a confusing or fuzzily worded message that, to some people, say absolutely nothing.

You do not want your people to read the organisational sustainability message and think 'what?'. And to make it worse, such messages can create a feeling that corporate sustainability is, at best, essentially greenwash and at worst, utter hypocrisy.

You may be thinking whether this means that we need to have two sets of messages? A conventional one for society generally and a separate set of guidance for people who need to use it to guide their daily decision making. Is this perhaps a way forward?

But a double message is tedious, confusing and possibly more expensive (despite the buy-one-get-one-free mentality in supermarkets, in management circles two usually cost more than one), it gives the bureaucrats more rules and confuses everyone else.

There are more elegant ways of dealing with the corporate sustainability message and also make it actually useful for everyone at work. But before we can deal with the message, we need to think about something more fundamental: what is the purpose of sustainability for our company?

The dictionaries provided a number of definitions for sustainability and various other definitions of sustainability are available to meet just about any preference. But a definition is not necessarily the same as a purpose. While a definition tells you *what something is*, a purpose can be seen as *the reason to do that something*. The purpose of sustainability is something many of us take for granted, probably because it seems quite obvious. A simple purpose of sustainability for a company can be:

> *Doing our best to sustain the parts of the company that are working well and eliminating or repairing the bits that aren't working so well.*

The example above is for a generic sustainability and should do well enough to keep a company going. But it is not precise, the definitions of 'well' and 'not so well' may be very different for different people, so it needs to become less ambiguous and more focused.

Once we established a purpose for corporate sustainability then we are ready to create a corporate message to support it.

KISS – Keep It Simple and Straightforward

There are lots of popular themes in society but 'keeping it simple' just seems not to resonate too well. Mobile phones have more functions than you can imagine and understanding the functions of video recorders has become an urban myth. Similarly, the sustainability message can be quite complex (see the Five Shared Principles above). That is not surprising, considering sustainability itself is really a very broad topic. Complex messages, however, often obscure a clear purpose and they are also less efficient.

Nevertheless, long and complex messages do have some strengths:

- First, they act as a barrier, thus protecting the people behind it (do you ever read the small print in the sales invoice or the user licence in your software?).

- Long messages also allow every nuance to be included, thus satisfying a large number of partisan groups and staying politically correct.

- Long messages makes a system look more complex, leading you to believe that the people who designed it thought long and hard over it.

- A long and detailed message leaves nothing to the imagination, so nobody can have a different interpretation (even though sustainability belongs to everyone and we are entitled to our own interpretations). While this may appear to limit dissent by defining everything, it is more likely to turn people off especially if they feel it is pointless to even try and seek a role for themselves.

- Long messages are boring, so maybe their creators are secretly wishing that you will be so bored you will go away and not bother them.

But deep down we all know this anyway, so why is it that we still create dull, wordy and, let us be honest here, fairly unhelpful sustainability slogans and messages?

A lot has been said about the benefits of keeping things simple.[7] My preference is:

> *Perfection is achieved, not when there is nothing more to add, but when there is nothing left to take away.*[8]

The Core Intent

What would be more effective in delivering sustainability in the workplace is perhaps a short and snappy message, a Core Intent statement, that can be used by everyone in the organisation during decision making and I do not mean something like London's 4R (see Figure 6.1).

---

7 John Maeda outlines this elegantly in *The Laws of Simplicity*, published by the MIT Press, 2007.

8 Antoine de Saint-Exupéry said this in *Wind, Sand and Stars*, 1939.

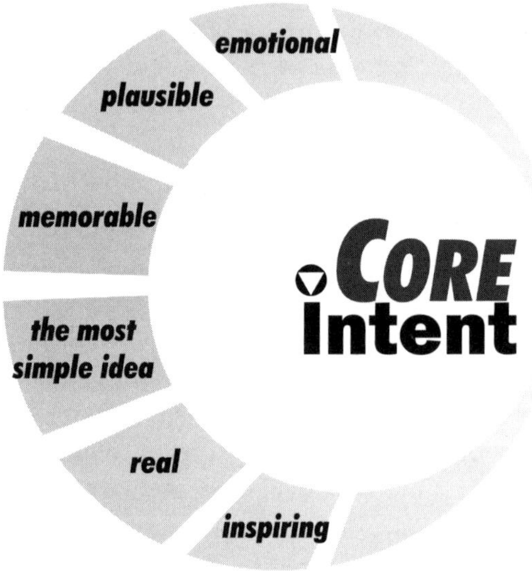

Figure 6.1 The Core Intent

The Core Intent[9] is what the company is about – what drives its beliefs and how it prefers to operate. The six components to make up an effective Core Intent are:

1. Emotional – it has to appeal to the heart.
2. Inspiring – it needs to outline the organisation's aspirations.
3. Plausible – it must demonstrate seriousness and commitment – although the intent itself needs not be serious.
4. Real – it must be a real aim that is related to the organisation's business;
5. Memorable – it has to stick in the mind.

9 I developed the Core Intent for this book as I cannot find any other pre-existing model that meets all the requirements. There are several similar 'intents': the US Army's 'command intent' and the 'strategic intent' of Hamal and Prahalad or the simplicity concept of the Heath brothers – all these inspired me to develop the Core Intent. (*Sources*: Marilyn Darling, 'Learning in the thick of it', Harvard Business Review, July 2005; Gary Hamal & C. K. Prahalad, Strategic Intent, ibid; and 'Made to Stick', Chip Heath and Dan Heath, Random House, 2007.)

6. The most simple idea – anything complicated will not help people make their decisions.

Or simply:

Table 6.2 The basis for a Core Intent statement

Why?	Because ...
Emotional and Inspiring	if it is not emotional and from the heart, people will find it harder to trust you; if it is not inspiring, they will not listen
Plausible and Real	if it is not plausible, people will ignore it; if it is not real, people will laugh at it
Memorable but yet very Simple in concept	if it does not stick in people's mind, then it is pointless; if it is too complicated, people will not remember

For example, the British Army uses the phrase 'Be the Best' as its recruitment slogan and British Airways used to call itself the 'World's Favourite Airline'. Both of these are also effective Core Intent statements because both allow every action to be judged against a common organisational standard:

- Will doing *this* help us to be: the 'best'/'world's favourite airline'?

Suppose a company choose: 'reduce waste in everything we do' as its Core Intent. Then every decision taken can be quickly measured against the Core Intent.

- Will doing *this* help us: 'reduce waste in everything we do'?

Other possible (but perhaps dull) Core Intents can include:

- Always consider reuse before disposal (dull Core Intents are, perhaps, better suited to team or departmental use where it can be used to focus the mind on a particular issue).

When a staff member is faced with a decision, they will interpret their own version of sustainability through the Core Intent. Ownership is suddenly passed to the staff member and since we all know the right thing to do anyway, we are now doing our thing within the organisation's perimeter set by the Core Intent.

A very good candidate for the industry's Core Intent statement is what John Ruffle of Allenbuild, suggests to me, 'professionals should be persuading their colleagues to consider sustainability in everything they do'. It does not get better than this, 'Consider sustainability in everything we do.' This is it – emotional and inspiring, plausible and real, sticks in your mind yet simple in concept.

You can ask yourself, 'will doing *this* has meant we have considered sustainability in everything we do?'

Besides John's brilliantly simple message, I have crafted a few other possible Core Intent statements, all of which can be readily checked during daily decision making:

- 'reduce waste in everything we do' – will working 'that way' help us *reduce waste in everything we do*?
- 'always putting health and safety first'– will 'that' Method Statement contribute to *always putting health and safety first*?
- 'at the cutting edge of architecture'– will choosing 'that' design put us *at the cutting edge of architecture*?
- 'financially prudent in all our undertakings'– will 'that' deal keep us *financially prudent in all our undertakings*?
- 'lowest cost construction company' – will 'that' new procedure make us the *lowest cost construction company*?

Generally a Core Intent statement will not have the thrill of the dynamic marketing slogans but it does help us in understanding what the company is about. Furthermore the Core Intent helps us go forward in a shared general direction because we understand the message and we can play our individual part within it. A simple and effective Core Intent can prevent the poor decision that ends up embarrassing both the organisation and the individual (see Figure 6.2). However, the most important reason for a Core Intent statement is that it allows empowerment of the employees within a corporate perimeter.

With one simple stroke, the concerns of many people in the organisation are dealt with:

- You can still practise your own version of sustainability but it is checked constantly against the corporate vision through the Core Intent.

- Innovations will not be ruled out since people can adapt their personal sustainability solutions within the Core Intent message.

- The control freaks will be happy too, since the workers will not be running the workplace – the Core Intent statement acts as the perimeter to limit empowerment.

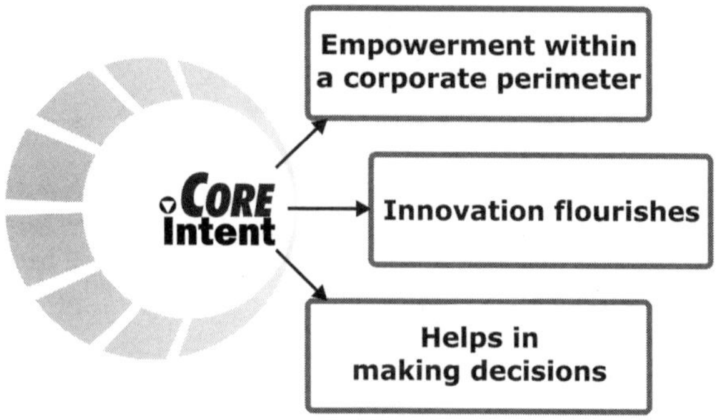

Figure 6.2 How the Core Intent empowers

The Core Intent, together with individual sustainability activities or policy elements, can combine to help produce a more balanced way of 'doing' sustainability. Every potential sustainability activity can be checked against the Core Intent for validity and then checked against the three sustainability components for balance.

However, the Core Intent is not a solution for all aspects of corporate communications. On its own, it is just a slogan. But backed with an effective framework of policies and activities, the Core Intent can act as a beacon in 'doing' sustainability (see Figure 6.3). Developing a relevant set of policies and activities are explored further in Chapter 17.

This is critical because none of us are sure what will happen and more importantly, how we and our organisation will respond. Interest rate changes; shortage of investment funds; volatile energy prices; changing government policies; new legislation; the skills shortage; resource availability; and so on.

The list stretches on and on. The problem is that we do not know what is coming next, only that something, probably unpleasant, is coming our way. A rallying point, such as an effective Core Intent, can free you mind from working out what is it you should worry about instead of delivering innovative solutions.

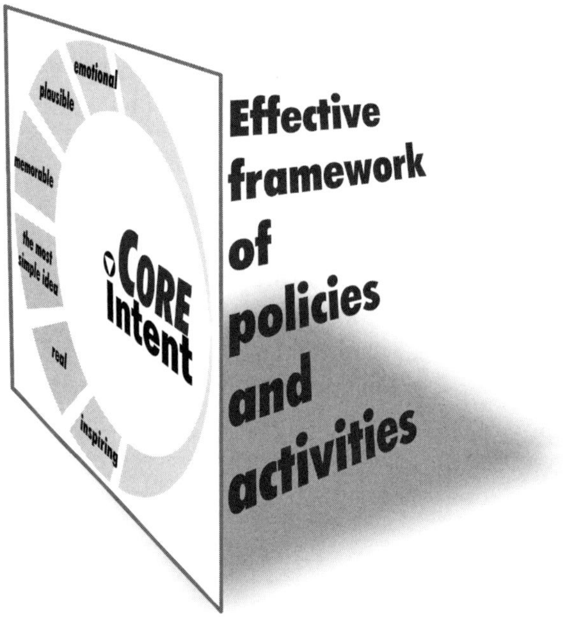

Figure 6.3 Supporting the Core Intent

In calmer times and in rougher ones, the sustainability of a construction company does rely on how well it manages its obligations towards the financial factors (earning income, managing costs and so on), the social factors (managing the shortage of skilled workers, dealing with construction site neighbours, health and safety at work and so on) and the environmental factors (legislation, self-regulating instruments like trade association commitments and so on). This ever-increasing load on the organisation coming from clients, neighbours, local and central governments can mean that current methods of working are becoming out of date.

In looking at sustainability and going forwards, backwards or sideways, we have encountered three factors that seem to come up time and time again, either as an roadblock or as something that can slow us down (see Figure 6.4). These are:

- money
- waste, and
- continuous improvement.

Since money is usually in short supply and then the other two: improvements and waste are the ways to go. My suggestion, and a Core Intent for doing sustainability, is:

reduce waste in everything we do and continuously improve on that.

I suggest this because it is a versatile option. You can safeguard profitability, support people and also reduce the corporate environmental impacts. It also allows us and our companies to become 'ready'. Ready for what? I have no idea but because I believe the world is changing at a pace that is almost faster than we can cope. But if our companies have low waste, we will have more resources available to deal with whatever comes with tomorrow, and if we improve on how we do things, then whatever comes, we will cope better.

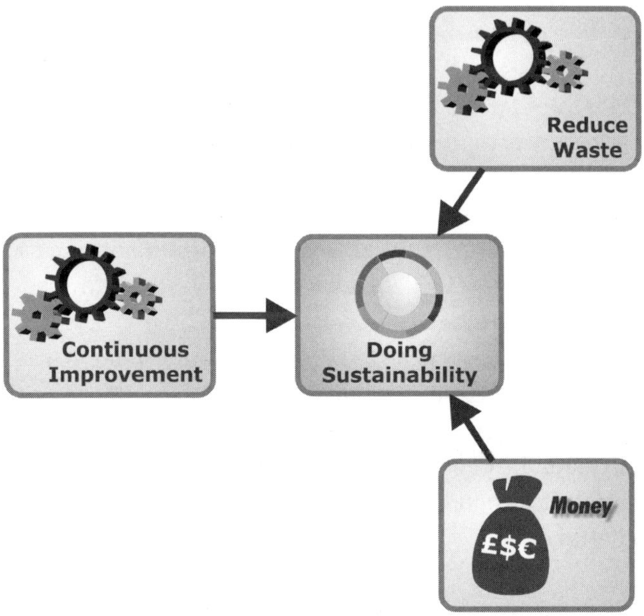

Figure 6.4 Reality check: sustainability drivers

Deploying the Core Intent statement is explored further in Chapters 16 and 17.

Communicating What You Have Been Doing – Marketing Sustainability

Doing sustainability is more than just working hard, it is also about effectiveness. Managing sustainability in construction includes doing sustainability activities – the day-to-day activities that reduce costs, engage people and reduce environmental impacts, and also carrying out some marketing. Marketing? I bet you thought sustainability is all about doing good, but the reality is that some of your efforts need to go towards telling people about what you are doing too – that is why PR is in the Criticality Analysis (see the Introduction to this book).

Telling people, marketing and promoting your ideas and results are important because:

- Your boss will learn about your ideas/successes and release resources towards your projects.

- Your clients will realise that working with your company boosts their reputation and so they want to work with your company more.

- Your competitors despair at your effectiveness and they either redouble their efforts (adding to the overall benefits for our society) or give up (you lose a competitor).

- Society is pleased with your efforts and the feel-good atmosphere makes doing business easier.

- Sustainability needs to get into the main stream of everyday corporate business operations and not in its own silo (or corporate ghetto).

Telling people about what we do is communications, but it is really marketing and when it is done on a corporate scale, it becomes Public Relations or PR.

Is PR a Vital Part of Doing Sustainability?

In a model of 'Value and Work' described later in Chapter 8, I suggest that the sustainability activities can be classified as a 'value-adding' activity, whereas the marketing or PR we do to show off our sustainability prowess probably contributes to our company's enhanced reputation. The PR is therefore seen as a 'value-enabling' activity as it enables the company to deliver value more effectively.

There is a moral dilemma in this. Although the impact of PR is quite profound nowadays, there is also a feeling that we do good mainly because we believe in it and not to provide ammunition for marketing. However, in our rather competitive industry climate, every company needs all the extra support it can get to convince clients of our capabilities. Therefore, humility and modesty often take a backseat to the need to promote our doing good – especially since the company has just paid for the sustainability activities (or not, as you will read later in Chapters 8 and 9).

The relationship between the actual doing good and the marketing of the same through PR is rather complex. Without any PR on our part, some of our sustainability activities may not happen because the decision makers will not have sufficient information to authorise the necessary resources. But without sustainability to provide the excitement, PR from the construction industry does not often offer a lot of differentiation between the various contractors (it is a different story with architects and designers).

I asked many of my colleagues whether they are driven by PR kudos or by doing good, the answer is often in two parts: '*we* do it for the right reasons' but 'others do it for PR or greenwash reasons only'.

Greenwash is often loathed by many sustainability professionals because it is considered to be insincere and often self-serving. Leaving aside the question whether a commercial company is there to do good in the first place, there is nothing wrong with being self-serving. In a competitive business environment, a company that does not look after its own interest is not likely to be around for long. However, using greenwash as a substitute to doing any meaningful sustainability action is really not a viable long-term strategy because a company's reputation will be tarnished by its persistent insincerity.

PR, however, is vital for spreading the word on sustainability. It allows the corporate executives to be tied into sustainability initiatives. It also encourages companies to out do each other in sustainability practices. And PR tells people – employees, customers, suppliers and the broader society that a company is doing something on sustainability. More cynically, good PR can help secure a (bigger) budget for you too.

And that, is why I put 'PR opportunity' into the Critical for Sustainability Analysis model!

7

Managing Sustainability: Your Personal Effectiveness

'Doing' sustainability is going to be part of our jobs for the rest of our professional lives. It may be called different things but essentially, we will be required to do something about the way we use resources: finance; people; and materials. This is no bad thing especially if we want to be able to enjoy our pensions in a pleasant environment. Making sustainability happen is something we can all do with little difficulty even in a recession, especially if we take a realistic approach towards what is readily achievable. In other words, we are probably not going to go for the UK government's high-style 'Shared Principles of Sustainable Development'. We are going for something a bit less elegant but perhaps more down to earth.

Managing sustainability requires deploying specific tools and procedures but all these come to nothing if one critical underpinning factor is not dealt with first. This is people: specifically you and your colleagues. What you do and how you do it matters, but it does not matter as much as how well you do it. Murray Coleman suggested sustainability should be at the core of an organisation and not as an appendage. Louise Perry, Group Sustainability Manager at Carillion, a builder, told me how Carillion started paying attention to sustainability since 1994 and over the years, integrated sustainability into overall business objectives. This ensured that all construction projects have detailed sustainability actions plans that implements the corporate strategy at a local level.

To do well, we need to start with sustainability at the personal level – your personal effectiveness. We will not be talking about how to save trees, but instead working more effectively and indirectly save trees. Similarly, we will not discuss how to get the most use out of a sheet of plasterboard, but about how to become more effective in dealing with waste problems. John Ruffle suggests we look at our own working environment and perhaps start doing sustainability from there.

Because sustainability is going to be a permanent feature in our jobs, we need to consider how we deal with it effectively. I suggest we need to take a professional and pragmatic approach to sustainability in construction, but what does a professional and pragmatic approach mean in an emotive subjective like sustainability?

Professional and Pragmatic

Being professional means that we treat it seriously, just like any other topic at work. We look at the problems, the issues and everything associated with it in a professional way. I suggest this because if we treat sustainability as a special case, and not as part of the daily work, we will soon have specialists running it. (Oh, wait! We have that already, oops!) The responsibility will then be taken away from everyone and passed to a special sustainability project or handled by a specialist team. Before you know it, sustainability can end up in a corporate ghetto and it becomes someone's specific responsibility and we are back to where we started: looking at the specialists and grumbling about what they do. Since sustainability belongs to all of us, therefore we will *all* take responsibility for it.

What About Being Pragmatic?

Being pragmatic is to accept that we are probably some way off from achieving a corporate sustainability transformation right now (see Chapter 17), and we are not going to get there in one easy leap either. Instead we need to think about the steps in our roadmap, how we will get through each stage successfully. To do that, we may need to change the way we think, the way we work and the way we deal with colleagues. Most of all, we probably need to remember that although each of us may only make a small contribution to sustainability through our individual efforts, together, our small individual blocks can turn into a pyramid of achievements.

Sometimes, however, we need to be flexible. Sometimes a favoured method may need adapting before it can be used effectively. So I am pragmatic in that I have adapted many tools in this book from their original application: heavy plant maintenance methods are adapted to help us measure how effective we are at work; and workplace organisation methods from a warehouse can be used to illustrate how we manage our information.

Being pragmatic is also about being realistic about what you can and cannot achieve: for example, for whatever reasons, sometimes you will not be able to get sustainability elements to balance equally. If there are financial obligations to fulfil, they will need to be dealt with first otherwise we may not have a business at all. You need to be professional and just get on with it. Just remember BATNEEC or Best Available Techniques Not Entailing Excessive Costs,[1] or do not go for things you cannot afford, but get the best with what you can afford. BATNEEC came from the Integrated Pollution Prevention and Control (IPPC) approach towards environmental management, as part of the Environmental Protection Act (1990) legislation. This is a dose of realism for some and unfortunately, a get-out clause for others.

The final important point about being pragmatic is that sustainability can often provides a range of solutions for every situation: these can range from 'just right' to 'acceptable' and including some with compromises. Compromising is often something that is inherent about sustainability, especially at an individual person's level.

For example:

- You want economic and social progress, it will use up some natural resources.
- You want a very low cost product, someone in a lower-wage economy will have to make it and it will need to be shipped here.
- You want the ultimate hazard and risk-free construction project, it will probably cost you more and take longer.

So the professional in us will have to accept the compromises and the lack of perfection pragmatically. Otherwise we may drive ourselves crazy without coming to terms with it.

1 BATNEEC is something the UK government emphasised in the early 1990s, this is explored in more detail in Chapters 13 and 14.

Your personal effectiveness in managing sustainability matters for several reasons:

- You need to set an example to others if you want to see sustainability become a reality at your workplace. It is no good relying on other people to kick-start it, you need to do it yourself.

- Before you can kick-start or contribute to the corporate sustainability drive, you need to make sure your day job is done effectively and efficiently. Otherwise, your boss will just tell you to get back to doing what they pay you for.

- Doing your day job well is crucial any time but even more so during a recession or a sluggish recovery, so we need to make ourselves more effective in everything we do.

So far we talked about personal effectiveness as though it comes naturally. Well, maybe for some people it does but for the rest of us, it takes some hard work and perhaps a technique or two.

Doing What Needs to be Done and Doing it Well

Personal effectiveness is about doing things well and using less resources and making less waste. This can be extended to looking at an individual's strengths and weaknesses. This aim is to maximise your effectiveness through understanding your strengths and roles.[2]

1. EFFECTIVENESS FROM YOUR STRENGTHS

Many of us are familiar with the SWOT analysis of Strengths/Weaknesses/Opportunities/Threats (see Figure 7.1). Clearly there is some link between the four elements: strengths allow you to exploit opportunities and they also let you neutralise threats. Weaknesses are exactly what they are – weaknesses.

Individually, we have strengths and weaknesses and would life not be easier if we can deploy our strengths in everything we do? After all, we are likely to achieve a lot more if we stick to doing things we are good at.

2 The approach I take in this book is not a particularly conventional one, since there are already lots of books taking the conventional approach based on skills and capabilities.

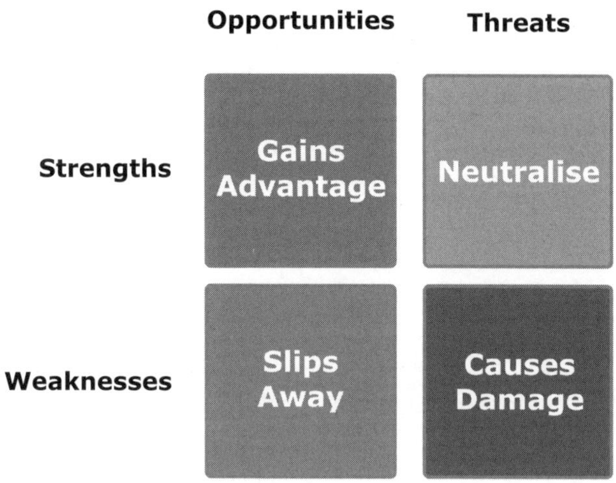

Figure 7.1 Effectiveness and strengths

In traditional management development thinking, we are always supposed to be improving our weaknesses. This approach was accepted generally as part of a professional's self-development – remember all the interview tips about preparing to mention three strengths and three weaknesses? However, recent thinking suggested that maybe we should be concentrating on developing our strengths instead.[3] This can be easily tested ask ourselves this question: will you hire or promote people on their weaknesses or on their strengths?

Often, fixing up weaknesses can seem a mug's game, it is like adapting and re-adapting an old building, it can be expensive in the long run and the benefits you get are not likely to be the same as a purpose-build unit. Ultimately you would probably do better to enhance your strengths and deliver more.

You may be thinking: wait a minute, I cannot go through life just developing my strengths. There are things known as appraisals where my weaknesses appear to be the only items of interest to my boss! Conventional thinking suggests that we always need to think of some weaknesses to mention during appraisals and job interviews.

3 Strength-based development was developed by Marcus Buckingham of the Gallup Organisation (the people who do polls). Both he and Tom Rath wrote DIY books on how to discover your strengths. In addition, there is also a management process called Appreciative Inquiry which focuses on the positive.

This is a moment of truth – you and your boss need to work out what is the most effective way for you to deliver value. I am fairly certain it is not via your weaknesses. The difficulty is that we are brought up with conventional thinking and talking about our strengths seems embarrassing. However, if we are to be successful, then we need to focus on what we are good at. This is really common sense – if you do what you are good at doing, then you are more likely to succeed.

There is an old saying:

> *If the only tool you have is a hammer, then every problem looks like a nail.*

This was to advise us about not being narrow-minded. Let us turn this saying around a bit:

> *If the only tool you have is a hammer and every problem looks like a nail, then you are onto a good thing!*

The trick, of course, is about turning every problem into a nail. To do that, you will need to discover your strengths first.

There are many ways to discover your strengths, and there are even websites that offer some sort of survey (reality check: if the survey is going to be any good, money needs to change hands first). A more long-winded way is to work out (by yourself) what are your strengths. This approach is based on you analysing the successes in your life and then working out your strengths. I have provided some basic guidance on the do-it-yourself version of finding your strengths in the accompanying CD, but you may wish to do the comprehensive web-version for the price of a book.[4] The expensive and more bespoke version is perhaps to go to an Appreciative Inquiry consultant[5] or a Belbin consultant.[6]

Once you know your strengths, you should be able to have a clear picture of how you achieve your successes. Everyone has a natural way of doing things

4 The same people (Marcus Buckingham and Tom Rath) who promote the strength-based development also wrote the books – both under £10 in 2011 from online bookstores.

5 Appreciate Inquiry (AI) is about focusing on the positive. Have a look at (www.aradford.co.uk) for more details.

6 Meredith Belbin and his team developed the concept of team roles in the 1970s. These have been refined since, have a look at (www.belbin.co.uk) or if you read Greek, try (www.belbin.gr)!

and if you stick to this, your chance of success becomes higher. Your natural way is your 'hammer'.

Having a strength-based definition for effectiveness is all very well, but it is a bit vague. My own strengths[7] are: strategic; achiever; 'relator'; 'intellection'; and 'maximiser'.[8] Never mind that some (or all) of these are make-words, what does it mean for me in everyday situations? How do I use 'intellection' (which means, I believe, thinking it through) in helping my colleagues achieve value? It may be more useful on a day-to-day basis to look at the things we actually do. The strengths analysis will tell you how you deal with situations, but we need to map the situations first.

Countering the above is a simple argument – if your job requires you to do something specific, then whether doing it requires a strength of yours or displays a weakness, you still need to get it done and do it well – whatever the management thinkers say. A professional and pragmatic manager will try to do well in all areas but prioritise the areas that will deliver the most value – both to the individual and the organisation.

2. EFFECTIVENESS FROM YOUR ROLES

To be effective, we need to know what is it we do everyday (see Figure 7.2). We do a lot of things, talking to people; go to meetings; write notes; move bits of timber; pour cement; climb steps; eat dinner; and so on.

But if we break these activities down to the most basic level, we basically 'do' just five unique activities:

1. **Observing**: looking; hearing; listening; feeling; tasting; smelling; and various other activities that collect information.

2. **Analysing**: making some sense or otherwise of the information we gathered.

3. **Communicating**: sending information to others and receiving information from others.

7 This is from the 'Strengthsfinder' web-based programme – look up the book of the same name by Tom Rath.

8 These bizarre make-words are all from 'Strengthsfinder'.

4. **Decision making**: using the observation; analysis; and communication to decide on a course of action.

5. **Doing**: carrying out that specific course of action.

In reality, these work roles interconnect with each other and we often are not aware that we move from one role to another. We see a sheet of flame which tell us a situation is not good, so we shout 'FIRE!' and decide to get out while the going is still good and walk swiftly towards the fire exit. Nevertheless, we may have different levels of effectiveness in each of these roles, for example: quicker in analysis; more thorough in observations; and so on.

Figure 7.2 The five basic work roles

Using my own analysis: 'intellection' from Strengthsfinder as an example, what does this mean in optimising my five work roles and trying to be more effective?

Table 7.1 Examples of the five work roles

Example: Applying 'Intellection' (or thinking it through) to Maximise Effectiveness	
Observing	Stick to what is there, do not over imagine what is not relevant or not there
Analysing	Go through possible scenarios of what caused it or what can happen next
Communicating	Don't get too esoteric unless I am talking to academic and/or specialist colleagues!
Decision Making	Check the decision for flaws, but don't get indecisive
Doing	Remembering that I am better at 'intellection' than actually doing!

With a checklist like this, I can now spot potential pitfalls that arise from my personal style and deal with them accordingly.

We all have different capabilities and we do better in some roles than others. The SWOT analysis, discussed earlier, tells us that if we play to our strengths, we can make the most of any opportunity that comes along. But what if we are weak in any of these five role areas? Since these roles are unavoidable, then the comment about not working on your weaknesses needs to be reviewed.

WORK ROLE 1: OBSERVING

This is something we do continuously, we also naturally filter out the details we feel do not enhance our lives (see Figure 7.3). The trick is fine tuning the filter so that unnecessary background is not logged but useful information is retained.

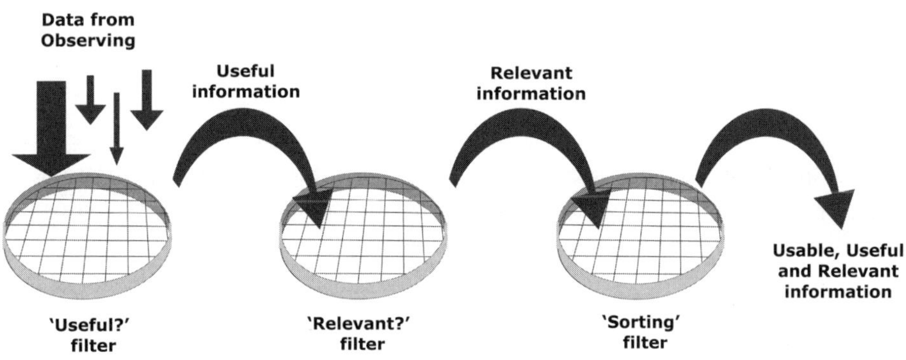

Figure 7.3 A cascade of observing

A simple filter is to adapt the SMART[9] target setting protocol.

- **S**pecific: log what you observed.
- **M**easurable: give an indication of the scale of the observation so that it has some perspective when compared to other observations.
- **A**ccurate: make sure you only log what you really observed and not what you think you observed.
- **R**elevant or realistic: does what you observed matter in your life/work?
- **T**ime: log the time of your observation, there may be a 'use by' date for what you observed.

WORK ROLE 2: ANALYSING

We all analyse what we observe, some of us are better at it than others. Analysis depends on what tools you have but more importantly, it depends on the quality of the observed information. This book contains a range of useful analysis tools to help you get the most out of your information (see Chapters 13 and 15, as well as the CD on tools).

WORK ROLE 3: COMMUNICATIONS

If we cannot communicate effectively, we will be dependent only on our own observations and analysis. While this may be totally adequate, but in most cases, having additional views tend to be helpful too in providing a bigger picture.[10]

Besides, you may have the best ideas about how to improve sustainability, but without effective communications, are you sure and who else will know about your ideas? Sustainability, like any other areas of business, relies on communications to make it work (see Chapter 6).

9 SMART targets are: Specific/Measurable/Achievable/Realistic/Timed – see Chapter 15.

10 This seems to be conflicting with the DMAIC tool described in Chapter 15 where it is emphasised that decisions should be based on facts and not hearsay. Nevertheless, in forming a view, sometimes opinions can add a valuable emotional viewpoint to a factual analysis.

Some people are natural communicators, they can talk irate bosses into giving them pay rises, get grumbling suppliers to do them favours and charm the clients every time. Naturally, we hate these people, but we are also a bit envious. Would it be good if we too can talk, write, present with the same effortless elegance? Today could be your lucky day because even if you are not a natural, you will find out how to train yourself to become a bit more natural, a little more elegant and use a lot less effort.[11]

Think about the times you fell under the spell of a master communicator, why do you feel so invigorated, so involved and so keen to do something? What is so exciting about these presentations, articles and conversations? Effective communications is not just about the content, but also how the content is delivered.[12] It is about how much you believe in what you are saying, how well you understood your information. It is also about paying attention to your audience's needs and going forward from there to fulfil them (see Figure 7.4).

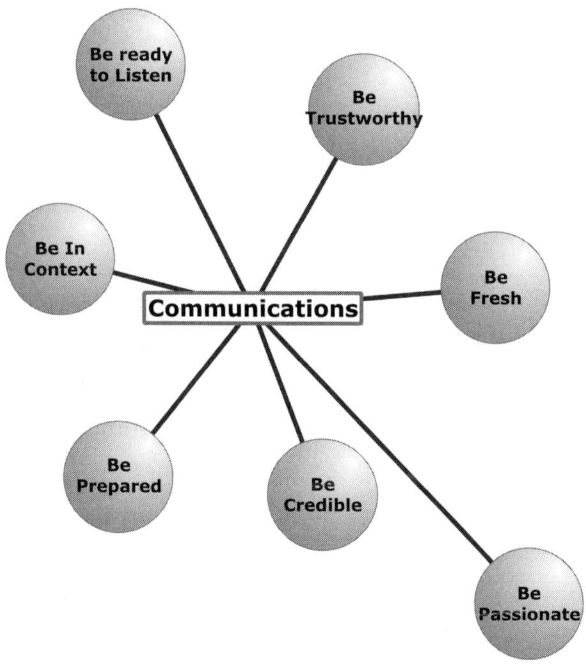

Figure 7.4 Effective communications

11 A very useful guide on getting communications right is *Ten Simple Secrets of the World's Top Business Communicators* by Carmine Gallo.

12 The other components that contribute to excitement are: context; and concept (see Chapter 14).

Content: to get your message accepted, it must be credible. Always check for errors and mistakes as they will reduce your credibility (see also Core Intent: real, Chapter 6).

Be in context: you may have valuable information, but if it is not in a form your audience can use, then your message is lost (see also Core Intent: plausible).

Preparation: you need to know your stuff and know it well, so you can have it ready to use at any time. Who knows when the next opportunity will come by?

Passion: if people do not think you care about what you are saying, then they will not pay attention. Your passion is what brings people to your ideas (see also Core Intent: emotional/ inspiring).

Trustworthy: information is power, using information ethically increases people's trust in you (see also Core Intent: real).

Fresh not dull: do not use the same message over and over again. It is boring and people will switch off. Reinvent, refresh, renew. If people want to hear the same thing over and over, they can always get a recording.

Be ready to listen: what does your audience want and can you give it to them? (this is explored in greater detail in Chapters 12 and 17).

WORK ROLE 4: DECISION MAKING

Although this sounds like what we do everyday, there is a bit more to it. Making a decision does not necessarily mean we need to carry it out (remember your New Year resolutions?). Decision making is an outcome of your observation, analysis and communications. Any of the three done poorly: wrong information; inadequate analysis; and mis-communication can all result in poor decision making.

Do you always make good decisions? There are hardly any checklists for making 'good decisions'. There are only more analysis tools. Nevertheless, any tool that helps with decision making should:

- help you clarify the benefits of the chosen action;
- highlight potential pitfalls;

- demonstrate whether you have the capabilities to carry out the chosen action (remember the lapsed New Year resolutions?); and
- show how your chosen action impact on other people.

A simple sustainability decision support tool is the criticality analysis, described in the Introduction.

WORK ROLE 5: DOING

This is both the moment of truth *and* the point of no return. Since just about everything we do will have some form of environmental impact, we return to the main theme of this book: whatever you do, do it with the minimal waste.

The outcome of any 'doing' is a combination of your knowledge, skills, experience and competence (see Figure 7.5). These four components are linked and the amount of value-added is related to how well these four complement each other.

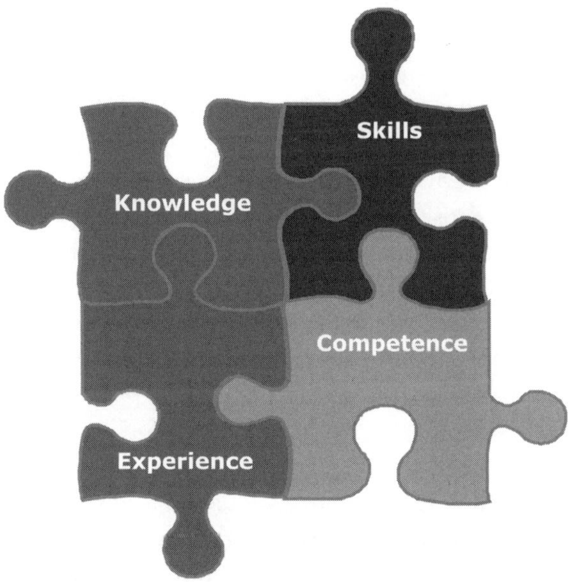

Figure 7.5 The four components of doing

- Knowledge is about your understanding of the chosen activity.
- Skills is about how good you are in carrying out the activity.
- Experience is about how familiar you are with that activity.
- Competence is about how good the outcomes are.

A combination of these makes up our capability in doing.

What happens if you are not good at any of the five work roles? Having suggested that enhancing weaknesses is not the best way to go, this is looking like a big hole in that argument. Actually, it is not. Because these five work roles are activities that affect every aspects of our lives – whether at work or not, we do need to enhance these somewhat if we are to be (more) effective in our lives.[13] So we need to enhance any work role weaknesses by learning a technique or work with colleagues who complements your strengths and provide cover in our weaker areas. One way to deal with this is to carry out a Belbin team role assessment on yourself and your team.

Team Roles and the Effective Sustainability Professional

How you deliver your effectiveness depends on your 'team role'.[14] Anastasia Marinopoulou from Progress Through, a consultancy, said that there are nine Belbin team roles and each of these have a profound impact on how an individual responds to situations and interacts with others (see Figure 7.6). Often, misunderstandings, accidents, errors, omissions and repetitive work occur because of incompatibilities between the work and the team structure.

The Belbin methodology helps individuals and teams to see how they work, as well as how their colleagues work. Although some sceptics may view personality classification as something similar to horoscopes, the Belbin roles do demonstrate how your personality can influence the way you tackle work. Basically, your team role gives you an indication of how you work naturally.

13 You can imagine these as one-dimensional performance issue (see Chapter 12), the better you are at them, the greater the benefits. On a more mundane level, it is like being able to file your tax online or setting up your mobile phone just right.

14 Also see footnote 6 in this chapter. Some of the names for the team roles are a bit unusual and may require a bit of intellection to figure out.

Figure 7.6 Managing sustainability: the Belbin Team Roles

Many sustainability professionals, especially those from a technical background, often display some '**Specialist**' traits. This means they tend to focus on a relatively narrow area (like all professionals) but can sometimes seem single-minded. Nevertheless, specialists are self-starting and very dedicated in what they do. They also provide rare knowledge (sounds just like you, right?).

Generally, we are not stuck with a single team role, most people have several team role styles, some more dominant than others.[15]

The full set of Belbin team roles are:

Monitor-evaluator: strategic; discerning; sees all options clearly and makes good judgement. Unfortunately, they can be a bit weak at inspiring

15 My dominant Belbin role is 'Shaper', supplemented by 'Plant' and 'Monitor-evaluator'. Mapping these to my strengths elements, such as 'intellection', is not altogether easy.

other people and so not always a natural leader, although the logical approach is very welcomed in teams.

Resource investigator: very enthusiastic; good communicator; and develops an extensive network. The flip side is that loses interest quickly and gets a bit over-optimistic when they do show interest.

Plant: comes up with imaginative and creative ideas; good at solving problems. But not too hot in explaining things and can ride a bit roughshod over minor issues.

Shaper: drives people onwards; does not mind pressure too much; and gets things done despite problems. However, shapers can be easily provoked and can also upset people along the way.

Coordinator: this is the typical chairperson; gets along with everyone; and delegates very well. This means some manipulation and gets others to do the work instead. (The original name for this role is 'chairman'.)

Teamworker: often a nice person to work with; listens a lot; readily gets on with people but not too decisive at critical times.

Implementer: this is the loyal and reliable company worker; gets things to happen. Often highly efficient but also occasionally a bit inflexible and not always too welcoming of new ways of doing things. (The original name for this role is 'company worker'.)

Completer finisher: the very conscientious colleague; and can be a perfectionist. But not too keen on delegating since nobody else will finish the job in exactly the 'right' way.

See a bit of yourself in all/many/some/few of those roles? Just like the horoscopes?

According to Marinopoulou, the point about the Belbin team roles is that you have to watch out for the weaker roles in your personal role profile and try to ensure that you are covered somehow in those areas as far as team work is concerned. The nine Belbin roles can combine to form a more or less 'perfect' team of personalities and innate capabilities, you need to think which roles are lacking in your current project and whether you or your colleagues can cover those gaps by performing additional roles. This clearly will not be easy since they are not going to be your natural roles but in any project, someone has to be the completer finisher!

Playing a role to make sure a job is done properly is not news to the sustainability professional or any professional. Most people tend to have to take on extra roles often in their work. We tend to fill in the other roles best we can. However, just like working in a messy workplace, we are less effective when the 'role' environment is not right, we do not always 'do' other roles very well.[16]

Most of us have to act as the 'Completer finisher' when a senior colleague or a boss dumps some work on us. Sometimes we have to become a 'Shaper' and crack the whip when we want to get things moving. We also need to scrounge for resources and be nice to people so that they will help us. Furthermore, we can also be very critical when we monitor and evaluate a proposal by a rival colleague. So we are doing it anyway, perhaps without conscious knowledge of our team roles and most probably not how to perform in those roles effectively. But once we have some knowledge of team roles, we can exploit them and make the most of our natural style and also assume other necessary roles in critical times. This brings us back to using our strengths to its utmost to get things done.

Imagine the following situations and see whether you agree with me about which team roles are required. Will you and your colleagues be able to step into the breach and perform the necessary roles? These scenarios assume that you are a sustainability champion.

Situation One

You are at a meeting with the supervisors of all the subcontractors on the project. You wanted to use the meeting to discuss waste problems: what they were and how they came about. However, one supervisor claimed that his team does not make any waste and the wastes in his work area are dropped by other contractors. The others look sourly at him. Most sat with their arms crossed and one of the others said he had training from Andrew Pears and the

16 In these situations, an old metaphor comes to mind, 'a fish out of water'.

Cafe Van so they really do know how to reduce waste as well as to segregate waste. The meeting is not going the way you wanted. It is time to play a role.

1. A 'Coordinator' will sooth everyone and gets people to agree to undertake various actions, although you are not sure whether these supervisors will follow through.

2. A 'Shaper' can get the message across forcefully and will not accept miserable responses. But will there be an argument and you may have to pull weight using the contract or say something sarcastic? The problem is that it can damage relations.

3. A 'Teamworker' will listen to what everyone says, smile and make nice to everyone. The supervisors may uncross their arms and probably nothing will happen. But are you sure harmony is better than getting something done?

Situation Two

Very poor weather has meant a delay to the project and everyone is getting a bit tense since your schedule is a tough one. Your subcontractors seem to have lost interest in sustainability and waste reduction, because everyone is thinking of rushing as fast as possible to catch up. You have a monthly subcontractor managers meeting coming up.

What role should you consider playing to remind them of their commitment to sustainability?

1. A 'Monitor-evaluator' will point out that rushing means waste and more mistakes, so by rushing they will actually work slower. This is technically correct but will it raise morale and gain you support?

2. A 'Coordinator' will relax everyone and say the right things about working together to reduce problems (and waste) and everyone need to play their part and make everyone feel good. But if this is the good–bad cop routine, who's going to be the bad cop in making it happen?

3. An 'Implementer' can tell everyone to stick to the plan and things will happen. After all, every project has hiccups and sticking to the plan will get you through, this time as in times before. Is this inspiring or will your subcontractor managers ignore you?

Situation Three

A supervisor came to see you about his team. They have received your sustainability messages so many times that they are bored to death by another poster of skips and a man holding a handful of money.

Which colleague will you call on to help develop a new and exciting message?

1. A 'Plant' can come up with a wonderful new approach towards waste reduction, but because it is so ultra-cool, it is unlikely the team (and everyone else on site) will get it.

2. A specialist will develop a fact-filled message that explains everything very clearly and in detail. Unfortunately the team is likely to be so overwhelmed by the heavy-going message that they will probably have headaches afterwards.

3. A 'Resource investigator' will contact her network of colleagues and cobble together an imaginative message that represents the best in the industry right now. But in a chat with a colleague at Costain, she thought of something even more interesting and now cannot help with disseminating the original message because she is working on a better version.

4. (And this is the correct answer.) Hire Uly Ma to revitalise your programme! (just kidding).

Situation Four

The waste reduction innovation scheme has worked wonders and you are receiving about 30 ideas every week. You need to get these sorted out, but you are busy. Wait, you have just been told several undergraduate placement students are available to do the leg work.

So who do you select?

1. A 'Completer finisher' will be patient in sorting through the ideas using a set of criteria you provided but he wants to see all 15 selected projects through too. This means you have to explain all 15 implementations to him in great detail. Suddenly you are very behind in your work.

2. A 'Monitor–evaluator' gets back to 75 per cent of the innovators and asks for more detailed explanations. His critical style of questioning is upsetting people and the stream of ideas risks drying up.

3. A 'Teamworker' goes to talk to all the teams and they all like her friendly and open style, and promise more even ideas. But then she tells you all 30 ideas are great and should all be implemented. You stare at her and start to explain the Pareto 80–20 rule again.

Situation Five

A chance meeting with your CEO gave you an opportunity to sell your waste reduction idea. Now fame (and hopefully fortune) beckons as the boss has decided to let you run with the idea.

You need to consider which roles to play to get the design team on board.

1. If you are a 'Shaper' you can express your vision and drive people to get on board, but risks upsetting those 'not invented here' types.

2. By adopting a 'Coordinator's' style, you can make everyone feel comfortable about your idea (it is really their idea too, of course). And you can delegate all the tasks since they all own the idea now (hopefully).

3. A 'Resource investigator', on the other hand, will search only for true enthusiasts and bring them on board. The project will go

smoothly, but then not everyone is involved and the impact will be less than what you promised the boss.

Situation Six

The workplace organisation programme using 5S[17] has been a success, the site is less messy and the accident rates from slips, trips and falls have come down. Now you are worried that people will become complacent and then returning to poor habits.

What is the best way you can maintain the momentum?

1. A 'Teamworker' role will let you get into the psyche of the people on site and understand the many reasons why there may be backsliding. But then what do you do?

2. You adopt a 'Complete finisher' attitude and doggedly push onwards. Since other people are generally unreliable, you do it all yourself and work very long hours to remind and chivvy people to keep up the momentum.

3. You decided a 'Plant' is the better option and come up with a new twist on the original initiative and use that to motivate people to keep it up. You also come up with a very complicated incentives programme to give rewards based on a formula that includes rubbish bags, photo evidence and numbers of worn out gloves every week.

Although these situations are made up, they do indicate the options we face everyday. Every Belbin team role has its place and offers both advantages and disadvantages. The value of team roles is that it is a team effort. If you are working on your own, you need to be aware that it is unlikely that you can cover all the team roles superbly.

While this may all be familiar ground, the need to enhance our personal effectiveness is critical as it will allow us to do sustainability better and tackle the anti-sustainability: waste in everything we do.

17 5S is a Lean construction technique to enhance the workplace – making it safer, more effective and more efficient to work in – see the accompanying CD for more details.

PART 2
The Anti-Sustainability

8

Introducing Waste: The Anti-Sustainability

Waste is the opposite of sustainability. Waste costs money, it demoralises people and it literally costs the earth. In construction, we make a lot of it, over 100 M tonnes worth every year. My approach to waste is broader than the conventional sustainability/environmental management view of waste – in other words, more than skips and bins, or reuse or recycling. Waste is an example of companies doing business poorly or individuals working ineffectively. Reducing waste is probably one of the most direct way of doing sustainability.

'Waste Not, Want Not'

Some old sayings never go out of date. If we are to have no waste, what would it mean in the construction industry of the twenty-first century? No waste in construction can result in:

- At least 10 per cent fewer vehicle journeys because apparently we traditionally over order by 10 per cent in the industry so we save on transport costs and cut the environmental impacts of that 10 per cent.
- Significant savings in energy from not having to make the now unnecessary materials.[1]

1 Since much of the material used in construction is made from roasting rock or melting sand – you figure out what that means for the energy bill!

- Saving in time – UK and US research suggested that anything up to 26 per cent of a working week is spent doing non-value adding activities.[2]

- Since we need less materials, pay for fewer vehicle journeys, save energy and save time, we must also be saving MONEY!

We discussed how money, waste and continuous improvement are three significant and linked factors in doing sustainability. In this section, we put the magnifying glass on waste as well as its relationship to daily operations and sustainability.

When I discussed capabilities for 'doing green' earlier in the book, I mentioned waste reduction. When I proposed the definition of sustainability, I also mentioned waste reduction. Waste is the single element that defines sustainability. Well, it is actually an *anti-sustainability*. If there is not any waste, there would be less need to talk about sustainability at all. If we do not waste resources (people, effort, ideas, materials, money) then there will be a lot less financial, social or environmental factors to worry about. From there, we can start envisaging cutting waste as a possible solution to 'doing sustainability'.

What is Waste?

Waste is one of those words that means different things to different people (see Figure 8.1). To many people in our industry, it is the stuff they throw into skips. Some people define it more specifically: only stuff for landfill is waste; recyclable materials are not necessarily waste. To some people, waste is what other people spend their time doing!

Although our description of waste depends on our outlook and circumstances, we can readily develop several possible definitions:

- Waste is anything that is discarded but can still be used to deliver value.

- Waste is any activity that does not add value.

2 This was carried out by ntl/Telewest in the UK (2006) and AOL/salary.com in the USA (2005), and it represents a generic level of waste. Anecdotal reports from the UK showed that 30 per cent of time spent on creating waste is not uncommon in industry.

- Waste is anything that the client does not want.
- Waste is difference between the final result and your efforts.
- Waste represents a source of funds to pay for sustainability activities.

Figure 8.1 What is waste?

Let us examine these definitions.

1. WASTE IS ANYTHING THAT IS DISCARDED BUT CAN STILL BE USED TO DELIVER VALUE

This is a definition that is often used by environmental managers and people who are focusing on reuse or recycling. There is nothing wrong with this definition, except that most users tend to look at it from the viewpoint of 'tangible' products, in other words, things you can slap a coat of paint on.

However, this definition is actually very profound if you use it to describe other ingredients for value creation: ideas; time; and effort are three things that come to mind. Many people leave their jobs for other companies because these three things are badly managed at their previous employers. Companies discard the ideas of their people by not listening to potential innovations, managers disrespect the efforts of their teams by making them carry out pointless activities and directors waste the time of their workforce through meaningless bureaucracy.

The waste in talent through missing out on ideas, effort and time are readily translated into lower morale and, ultimately, lower profitability from lowered productivity and using resources to do unnecessary work. This definition of waste accounted for all three sustainability elements: financial; social; and environmental.

This first definition of waste showed that it covers a lot more than just what we put into the skips and bins.

2. WASTE IS ANY ACTIVITY THAT DOES NOT ADD VALUE

This is a fairly absolute description and has no ambiguity. However, it is not entirely accurate because it is a bit too simplified. Toyota Motors, an acknowledged expert in managing waste, shows how waste, values and work (see Figure 8.2) linked in a rather unique relationship:

- Work is the sum of value-adding activities, value-enabling activities and waste.
- Work is what we do when we are, well, at work – our efforts.
- Value-adding activities are what the client wants from us. They are the outcomes of our work: a building; a wall; a concrete slab; a master plan for urban regeneration; and so on.
- Value-enabling activities are what we do to make the value-adding activities possible. So it can be the welfare facilities on site,[3] the ConstructionSkills levy rebate for training, or the health and safety tool box talks that contribute to a safer workplace.
- Waste is: everything else!

3 A summer job I once had in a Far Eastern rolling mill had an interesting welfare policy: you get a free lunch brought to you at your workstation. Sounds good until you realise that in exchange for a box of rice and sundries, you lost your hour break for lunch and they gained about 50 extra minutes of work.

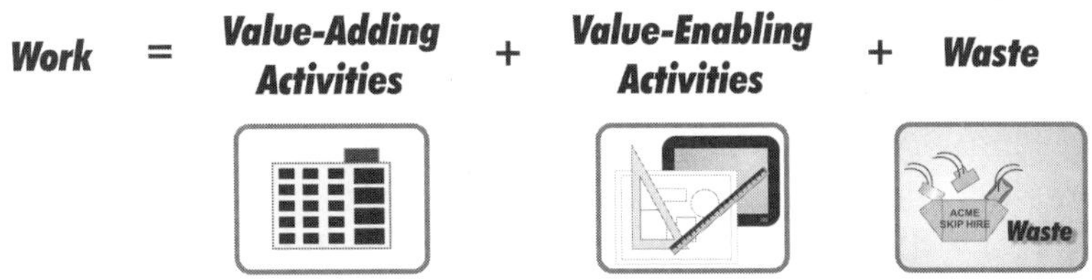

Figure 8.2 The links between work, value and waste

This definition means that we should think about work and waste in a different way. When we look at this relationship from a client's viewpoint, the entire equation becomes just that little bit more embarrassing (see Figure 8.3). Are clients supposed to pay for *our* waste?

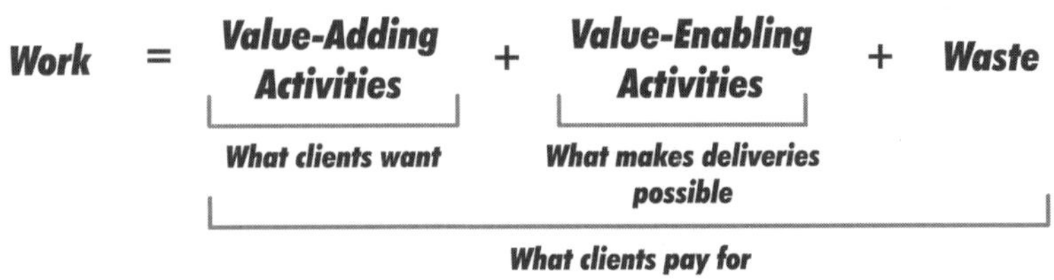

Figure 8.3 What clients pay for

In fact, this is made clear in the Site Waste Management Plan approach (see Chapter 10), but it seems that everyone have taken it at face value and see it as another box to tick rather than question whether it is reasonable to pay for other people's waste.[4]

Value-enabling activities form the framework of your company that allows contracted work to be delivered to the client. This can be a detailed health and safety process, a design function, a canteen, dry rooms and lockers. A pleasant reception area does not really add value to the deliverables for clients but it sure makes a better impression on clients. Since value-enhancing activities are anything that allows us to deliver the client requirements, then these are important activities to bear in mind.

4 Generally, we always pay for some waste in everything we buy. The difference in this case is that the SWMP makes it clear how much materials waste is there and so we actually know how much money is going straight down the toilet.

3. WASTE IS ANYTHING THAT THE CLIENT DOES NOT WANT

At first glance, this is exactly the same as the previous definition. However, there are subtle differences. Whereas we understand value-adding in terms of a building, a wall, a master plan, we may need to think a little deeper about value-enabling. What is value-enabling?

Welfare amenities and corporate responsibility are things that do not add to the functionality of a building, which is what the client wants you to build. But the wise client also realises that although these activities do not deliver functionality, they make constructing the required building that bit easier and smoother, so these value-enabling activities are a form of business cost that the client willingly pays for.

There is, however, a difficulty about value-enhancing activities. Just as for wastes, we all know when we buy a bucket or a bag of cement, we are also paying for the coffee the design team drank when they were designing the bucket or cement bag logo and we also pay for the paint job on the delivery vehicles of the bucket or cement companies. Neither of these actually make our buckets sturdier or more leak proof or our cement stronger, nevertheless, these costs are accepted generally. Nobody really thinks about them because they are hidden from us and we think they are reasonable costs. No one begrudges reasonable costs, such as a nurse on site or high-powered printers for construction plans and drawings. Nevertheless, as these activities themselves do not generate income, there is a need to differentiate them from the actual value-adding activities or unwanted activities.

It is, therefore, prudent to optimise these activities such that they deliver the best value for money. If you cannot optimise them (or are unwilling to do so), then it is best to make sure your clients do not know about the design team's away day to a luxury hotel in Paris, which they also paid for.

Unwanted work is over-design, overdoing, unnecessary transport and anything the client decides as not adding value (or not adding enough value to justify the costs!).

4. WASTE IS DIFFERENCE BETWEEN THE FINAL RESULT AND YOUR EFFORTS

This is another classical Toyota definition of waste.[5] It is clear, in black and white with no grey areas of ambiguity. It is equally applicable to the individual and to the team. You can test this readily at the end of each day by looking at all your achievements of the day and how much time these consumed. Your analysis is likely to find that 'effort' is rarely completely translated into the final results.

What makes this definition particularly interesting to us is that it seems to put a very definite split between our efforts and waste. It is actually like a wall, dividing valuable efforts, both creating and enabling, on one side and the bad news on the other. This description of the separation of waste and effort is something we do. When we get to work, we just get on with doing whatever we need to do, often subconsciously, to get the job done. This happens everyday, virtually everywhere. What we do not always notice, unless they are particularly galling, are the small things we also have to do to get round the problems as well as dealing with the actual problems themselves. We see these problems day in and day out, so much so that they become part of the background to our work. As they become part of the workplace, we do not notice them anymore. It is almost like the hypothetical wall is separating two workplaces: one which we get things done; and the other is where all the problems live.

Toyota has a name for this second workplace, they call it the Hidden Factory. This Hidden Workplace, be it a head office, a design bureau, a warehouse or a construction site, is part of the reality of working for most of us (see Figure 8.4). According to Toyota, most of us work round the Hidden Workplace, taking more time and using more resources to get the same job done.

5 As described by Taiichi Ohno, the 'father' of Lean in his book, *Toyota Production System: Beyond Large-Scale Production*, published by Productivity Press.

Figure 8.4 The Hidden Construction Site

I have one last definition of waste before taking a tour of the Hidden Workplace.

5. WASTE REPRESENTS A SOURCE OF FUNDS TO PAY FOR SUSTAINABILITY ACTIVITIES

At first, this sounds like back-to-front thinking. But let us examine this logically:

> Fact 1: sustainability has a cost. Currently, we need to do additional things to make sustainability happen compared to 'business-as-usual'.[6]
>
> Fact 2: companies do not like to spend money.
>
> Fact 3: waste costs money too – since 10–20 per cent[7] of materials coming on site can end up in the skip.
>
> Fact 4: the amount of time we spent dealing with the 10–20 per cent materials (destined for the skip) cannot be spend generating more added value for our clients.

6 Perhaps more accurately, 'business as before'.
7 Anecdotal evidence as well as estimates by Construction Clients Federation in 2003 provided this range of estimates, whatever it is, it is a lot.

Recently at my NoWaste project at the big Stratford project where we used a generic national figure of 19 per cent waste in our training programme. One operative remarked, 'so every fifth pallet coming in goes straight out in waste bins'. Exactly, not only are construction companies paying for that fifth pallet of materials – unused, they then pay their teams to handle the materials from pallet to bins through a series of movements. And that is not all, they are also paying for the waste to be hauled away, and then paying Landfill Tax or waste handling fees on top. I think companies may actually do better by giving the same amounts of money to charities, at least that is doing some good.

If we can recover the money currently lost to waste, then some of it can be allocated towards doing sustainability. It is unlikely that any company will spend all its savings from waste reduction on sustainability, but it is up to the sustainability champions to sell this message. Companies may respond more positively to a 'solution' that shows sustainability as a profit centre, than when sustainability is seen as a cost centre.[8]

By adopting this approach, the company does not actually need to spend 'new' or extra money for sustainability since it is using money that has been recovered from waste. It is kind of like reusing and recycling, but about money instead.

Figure 8.5 shows how this can work. In the business-as-before case, the company has a cost of sales which includes the waste component. By reducing waste, the cost of sales is reduced with a corresponding rise in profits. A portion of the profits can then be allocated towards doing sustainability. The profit margin remains higher than the business-as-usual case and yet a budget for sustainability is created without requiring additional funds.

In reality, the situation is a bit more complex. There needs to be accurate accounting of cost savings as well as a realistic projection of waste costs in future years from the business-as-usual case. But most critical of all is a willingness to accept that the cost recovery from waste to paying for sustainability will lag the initial investment in sustainability activities. This is where many companies show their reticence when faced with the necessity of paying for something up front and accepting that cost recovery will come in due course. In many ways, it is a reflection of the short-term thinking that is common in many areas of our industry. But then, you know all about that anyway.

8 And since sustainability is also a legally mandated cost centre, a natural reaction can be to just do the minimum, and that benefits fewer people.

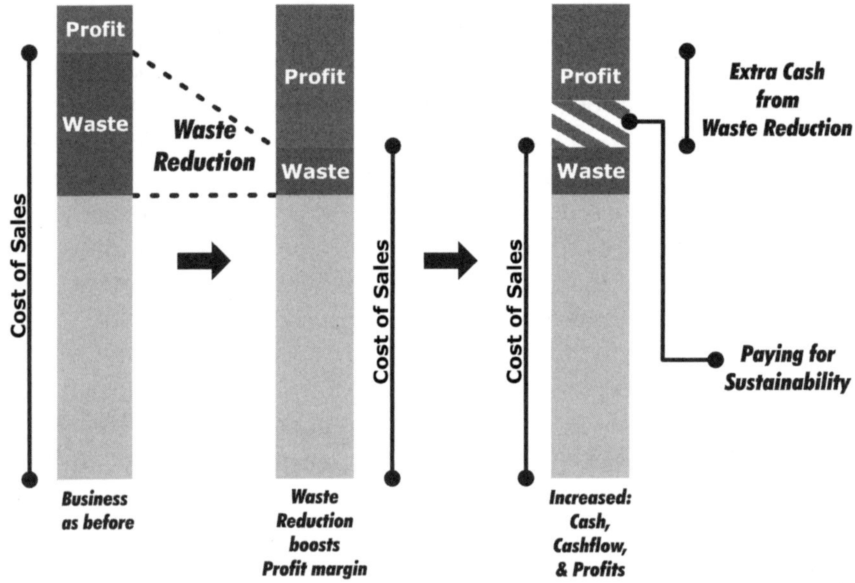

Figure 8.5 Waste reduction pays for sustainability

The good thing about waste reduction is that it is something everyone can take part in. It is also something, as Murray Coleman said, that allows people to relate directly to sustainability.

The Hidden Workplace

Toyota identified seven main wastes in their factories. Just about everybody else noticed that there are eight (see Figure 8.6).

The extra item is about wasting the talent and ideas from the people working at the company.

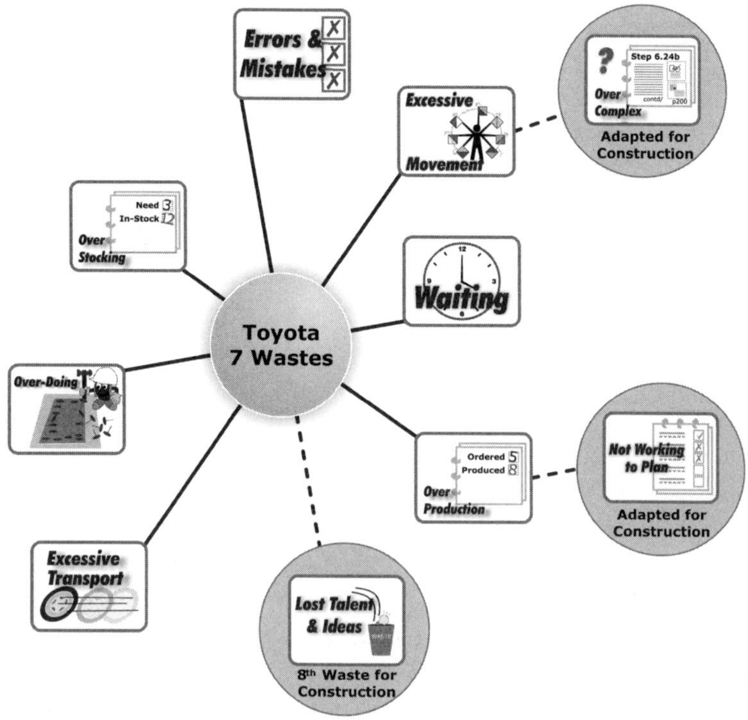

Figure 8.6 The Hidden Wastes

The full list of Eight Wastes, adapted[9] for construction are:

1. waiting
2. over-complex procedures
3. not working to plan
4. overdoing or redoing
5. excessive transport

9 Toyota's wastes are not necessarily the right ones for a construction industry, 'over production', a Toyota original, is something that happens in mass manufacturing but not in construction – we do not produce an extra building or additional doors and lift cores as 'buffer stock'. Our overdoing is more like redoing a job following changes to the plans.

6. over stocking

7. defects, mistakes and errors; and

8. lost ideas and innovations.

These are the hassles we all face in construction. We also know that although the Eight Wastes listed seem fairly high level, the results often mean more filled skips and bins and with it higher waste removal costs and Landfill Tax levies.

These wastes are also true for the design office, the sales office or the head office, whether for construction, property development, off-site manufacturing or service provision.

Although these wastes are a part of our work, we tended not to notice them because most do not appear to be directly connected with waste bins or come with green recycle labels. And these are the nastiest wastes because conventional environmental management ways of dealing with wastes cannot sort them out. The impacts of these wastes, however, are quite conventional: triple hits on the financial, social and environmental elements of sustainability. In fact, as we said before, waste is the anti-sustainability. Now let us pay a visit to the Hidden Workplace and see why this is so.

Waiting

This is something that happens to everyone. While you are hanging around, resources like electricity and water are being paid but not used and people also get bored. Your plant and equipment are also idle when you are waiting and you cannot get any productivity out of any of these already paid-for or hired resources.

We also know from personal experience that when people get bored, they become more careless and are likely to make mistakes and have accidents. This is a triple hit in sustainability terms:

- financial – costs money;
- social – demotivating and increases the risks of an accident; and
- environmental – wastes resources.

And what happens after waiting? Rushing to catch up – to get the job done. Rushing leads to mistakes, errors, greater risks of accidents and potentially higher levels of waste.

Over-Complex Procedures

When designs and procedures are too complicated, people tend to make mistakes or take longer to finish. This means the schedule has slipped and again people rush to catch up.

Why do we have procedures that are over complicated? Can it be that they are not right for this particular job? We need to make sure that procedures and methods are right for each specific job. Sure it seems cheaper just to bring the entire methods and procedures from a previous job to the current job, but it is never the same job. Using inappropriate procedures can be difficult and costly in terms of mistakes, time and safety.

Redesigning complex procedures by simplifying them can help us reduce waste by simply designing them out in the first place.

Not Working to Plan

How often have we had to stop what we are doing and change to doing something else? Changes are a big problem in a construction project. There is additional clutter from having another set of plans, drawings, materials or semi-finished components at the workplace. The increased clutter can result in misplaced information and materials among the growing pile; errors and mistakes as well as contributing to additional tripping hazards on a construction site.

The problems caused by from materials and teams arriving on site earlier or later often lead to disruptions. Results are: lost time; effort; and information as well as a lot of wasted resources like energy; water; and materials. These all end up costing more money.

Overdoing[10] or Redoing

Sometimes, despite our best intentions, change (and other things!) happens and we need to do the job again. We deliver what is asked for again (see Figure 8.7).

Figure 8.7 The reality of overdoing
Source: NoWaste programme

Changes seemed to be a way of life in the construction sector. Plans are changed, drawings are changed, schedules are changed – and not always communicated effectively to the teams doing the work. With the result being one team carried on doing what it thought were the right tasks, while the other team expected the tasks being done the new way. Poor communications is often the cause of this with everyone rushing to get their tasks done, it is not surprising that some of these communications slip through the net.

However, poor communications which resulted in overdoing is not necessarily the worse case. Because of poor scheduling, sometimes teams will be carrying out activities that *they know will need to be redone later.* I observed in many projects that the dry lining team were working even when the windows or roof have not been sealed or completed. Then comes the rain and the plasterboard ends up ruined and has to be redone. The same is true for joinery teams – they put up doors and handrails before the other components

10 Overdoing does not mean cutting corners, especially when it comes to safety precautions or environmental awareness.

are finished knowing they will need to come back to sort out the scratches and damages later. Painters despair of finishing before other teams because they know there will be rework: overdoing again and again.[11]

Excessive Transport

This can be as simple as moving things more than necessary to get the job done. These things can range from:

- **Information**: too much checking back and forth can mean the message itself is submerged; or too many layers of communications can dilute or even distort a message.
- **Materials**: moving heavy materials back and forth uses all kinds of resources, such as people, time, money, storage space, etc. Too much moving back and forth can contribute to accidental damage and even losing the item.
- **Semi-finished components**: these can half-completed reports, plans, building structures, and unless they are designed to be transported while half completed, you run the risk of losing vital components whether they are sheets of paper or window frames.

What causes excessive transport? Generally it is a mix of poor communications, poor design and poor training. It often happens when people did not get the right information; did not know what to do; or did not know how to do it properly.

IS RECYCLING OVER TRANSPORT? WELL, LET'S HAVE A LOOK

Over transport #1: some of the materials are delivered but not always necessarily used (perhaps over ordered, or damaged through transport/ storage, or discarded from error or mistakes).

Over transport #2: they are then binned or skipped and collected.

11 I asked some painters about putting warning signs like 'wet paint' to advise others working in the area and was told that it actually made the situation worse as people, being perverse, then touched the paint to check whether it is really wet!

Over transport #3: the waste material is consolidated and delivered to the recycling factory.

Delivering the remade material to the next user is not over transport as it becomes part of the raw materials supply stream. (But it still wastes resources: effort; and money.)

By being more careful in planning, ordering and using the materials, we can cut down on the waste, both in money and energy terms. Nevertheless, it is unrealistic to expect no recycling because some off-cuts are not reusable in practical terms.

Over Stocking

This happens everywhere, whether at home or at work. We are enticed by bargains or sometimes too lazy to maintain a regular monitoring of stocks. A good deal at the stationary suppliers means our stationary cupboard is bulging with photocopier paper, or the construction operative taking five boxes of fasteners to a job when three boxes will probably do. This is especially relevant if the stores are a distance from where you are working.

Over stocking carries penalties since it costs money to carry the excess inventory or stock. This diverts the money from other uses, and means you are tying up your cash unnecessarily. Once you bought all the materials, you also have to find a place to keep them secure from spoilage or damage, resulting in even higher costs. The problems do not stop there: for example, having piles of photocopying paper lying around can cause a tripping hazard as well as being a fire hazard.

Over stocking often leads to waste. If there are loads of materials lying around, the need to conserve or be careful will appear less attractive and people may be less careful with how the materials are used. If you have taken more fasteners than you need, you may not be bothered to return the extra boxes of fasteners to the stores. And it costs more money to haul the skips away full of usable materials.

Socially, it leads to poor working habits: if you know that there are 200 more boxes of fasteners in the stores, you may be a bit more careless in using them. Result: more waste.

The sustainability penalty also includes the wasted materials themselves and the energy and resources used to produce and deliver them.

Defects, Mistakes and Errors

Nobody is perfect and everyone makes mistakes. In fact, for many of us, learning from our mistakes is often one way to go forward. But mistakes and defects often result from poor communication, design and training.

Defects, mistakes and errors are the most obvious anti-sustainability elements. If we make a mistake, we have to do it again. So more materials are used, more effort deployed and more time taken up.

But wait, there is more! Our working environments have overhead costs: the value enablers, such as lighting, hot water, heating and ventilation.[12] These costs are generally accepted since they enable work to be carried out. However, when the work carried out is waste, or result in defects or rework, then the productivity of the overhead elements is lost as waste as well.

Making mistakes is not something anyone enjoys and it certainly lowers motivation especially when we believe the mistake is due to poor communications or inadequate tools (in other words, not our fault!).

Lost Ideas and Innovations

This is possibly the most serious waste of all. Since your competitors are likely to offer more or less the same package as you, how will your company differentiate yourself? You and your organisation are selected by the client because they have something special to offer. One way to gain a competitive edge is to be seen as an innovating company. This works best when everyone is contributing ideas for improvement.

Everyone has something to contribute because we are all unique – with different ways of thinking, doing things, we have different backgrounds and learn and assess situations differently. With such diversity, it is not surprising

12 Of course, on many construction projects, heating and cooling are free, the former in the summer and the latter in winter.

that everyone has something to contribute towards making sustainability happen at the workplace.

The difficulty with coming up with ideas in many workplaces is not the idea itself, but what to do with it. Some organisations establish channels where ideas can be assessed and then implemented, but many others do not. We have all seen suggestion boxes that we believe are really the front end to a waste bin. How can we encourage our colleagues to contribute ideas and suggestions if our organisation does not recognise or respect their contributions?

Besides from getting people to come up with ideas, the next problem is the way the ideas are processed. Without an open, responsive and respectful process, it is unlikely that any initial enthusiasm will be sustained.

Encouraging and sustaining a culture of innovation (which is what this is about) is something that is a critical part of any sustainability initiative. How to get people to change their behaviour, innovate and continually improve is described in more detail in Chapter 16.

9

Dealing with Waste

Conventional waste management systems and tools used in construction are often effective up to a point. They are effective for tangible or material wastes – the stuff that ends up in skips and bins. They are not too useful when we wanted to deal with the Eight Wastes from the Hidden Construction Site. This is probably because conventional waste management often comes from environmental management instead of business management (see Figure 9.1). Therefore they are good for 'end-of-pipe' solution when the physical waste is already generated. The focus on reuse and recycle is very good for timber or sheet metal, but fairly useless for waiting or lost talent and ideas. This means the Waste Hierarchy can be further divided into all wastes and tangible wastes.

Conventional thinking also relegates designing out waste to the designers and architects rather than considering that construction workers can also design out waste from their processes and procedures.

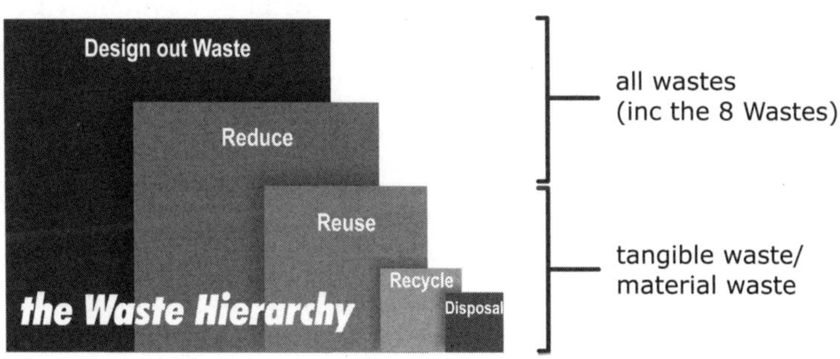

Figure 9.1 A different way of looking at waste

The majority of the Eight Waste elements are not readily recoverable once they happened, although the final outcomes of these wastes appear in our skips. The Eight Wastes are really things that cause the bins and skips to be filled. There is a similarity with electricity: you cannot see it and once you have used it (whether it is to provide light or to drive a motor), generally it is gone for good. So if the lights are left on in an empty office, or a motor left running without any load, the waste happens but we do not always notice them.

The Eight Wastes are about as tangible as electricity but more difficult to measure. There is no equivalent 'waiting waste' meter like the electricity meter to record the amount of usage. Jon de Souza of Constructing Excellence, a sector body, believes that we need to take waste more seriously and integrate waste reduction into design and construction. It is certainly a more valuable way for a company to operate than to sort out waste as an end-of-pipe issue. Perhaps another reason why the Eight Wastes are not readily addressed by environmental or sustainability management is because they tend not to be the actual contents in skips and bins.[1] The Eight Wastes approach to waste is process or operations focused and quite different to the environmental science approach deployed by people most closely associated with sustainability in our industry.

Dealing with the Eight Wastes is not clean cut like specifying FSC[2] approved timber or buying renewable energy. To purchase or FSC timber or green electricity, a manager needs only to justify it on costs, the do-good value and sustainability kudos. Several phone calls later, it is more or less done. However, to reduce 'waiting' is much more substantial. For those familiar with working through legislation or whole life cycles, it may not even be clear where to start looking at reducing the waste from waiting. Calculating the carbon footprint of waiting? The task can be overwhelming. And it is overwhelming because conventional views of sustainability often show a silo-like mentality with the environmental management team forced to deal with the day-to-day activities in isolation. But since sustainability is about managing resources effectively there needs to be a more substantial inclusion of all the company's operations. Much of the well-tested techniques from operations and quality management can be adapted to deliver the upstream gains and complement the downstream reuse and recycling activities.

---

1 The other reason why the Eight Wastes do not figure on the conventional sustainability radar is that it is from Lean operations – not something this industry is all that familiar with.

2 The Forest Stewardship Council certifies forests, among other activities, to ensure that people buying wood products that are from well-managed forests.

Quality and process management generally focus on reducing variation and increasing the efficiency of the process.[3] These targets are entirely compatible with sustainability objectives (see Figure 9.2).

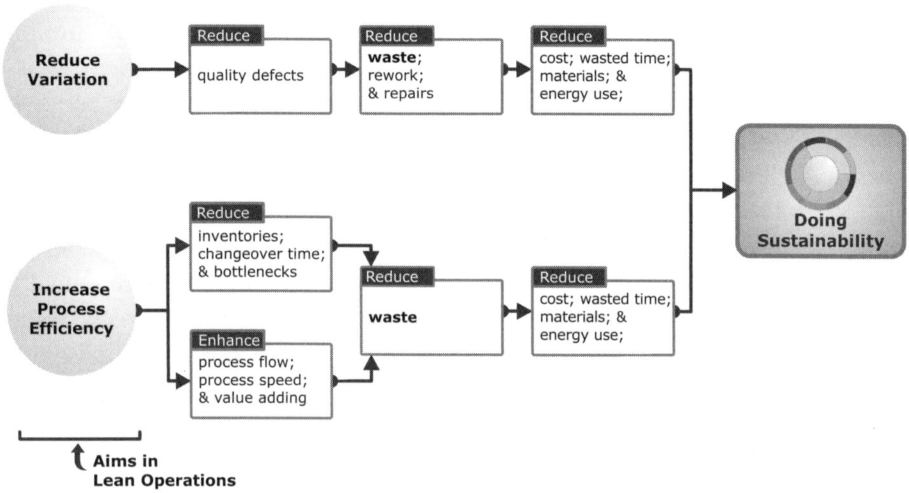

Figure 9.2 From quality and process management to doing sustainability

The Anti-Sustainability

I have suggested that waste is an anti-sustainability, this then offer us a simple path towards 'doing' sustainability. Simply by reducing waste, we will be 'doing' sustainability. Is that it?

MAKE SUSTAINABILITY HAPPEN BY REDUCING WASTE[4]

Clearly it cannot be that simple, otherwise everybody and every company will be doing it. Perhaps we should rephrase it: is it really that easy?

But since our industry creates about 100 M tonnes of waste each year and sends every fifth or sixth pallet of ordered materials to the skip, then it is evident

3 Nancy R. Tague's book, *The Quality Toolbox*, is published by the American Quality Society. This shows the various tools and techniques to enhance process efficiency and reduce resource wastes.

4 Although these six words more or less say it all, alas my publishers reckoned no one will buy a book with just six words in it, hence the other 80,000 plus words!

that, on the whole, we as an industry have not really cracked the problem. Reducing waste to achieve sustainability therefore appears neither simple nor easy. Furthermore, people are starting to realise that actions like switching off lights and not printing emails on paper are not ever going to be enough. Maybe it is because the Eight Wastes are so embedded into the way we work that they are hidden from us. If these wastes are hidden, how can we reduce them?

Also, it needs to be made clear that waste reduction is not the complete sustainability solution. Sustainability covers a lot more than waste reduction even though cutting waste can solve a lot of sustainability issues. However, for non-specialists or 'the rest of us', this is something everyone can do and where we can all make a difference. What is more important is that during a combined recession, sluggish recovery and credit crunch, waste reduction is probably the most effective approach towards 'doing' sustainability. Develop the necessary waste reduction discipline and your company will be doing sustainability, not necessarily the ultimate in sustainability but still a good effort. It is also a very good approach after the recession too, releasing resources tied up by wastes and generating cash – you just can not beat that – in good times or bad!

Dealing With the Eight Wastes – An Operational Approach

The Eight Wastes are operational wastes and because every organisation has an operational element, these are universal and ever-present wastes. My experiences showed me that just about every contractor suffer form these Eight Wastes to a greater or lesser extent. Since these wastes come from people doing their work, we ought therefore to look for operational solutions to tackle them. There are many methods in industry about dealing with wastes and problems. Two effective ones are Lean operations and Six Sigma Quality. Both these techniques can be adapted to suit the needs of the construction industry and sustainability.

In general, both these techniques are compatible with sustainability, albeit with one major exception. Both Six Sigma Quality and Lean operations are focused on the workplace and its primary external links such as the customers and the supply chain. The society and its needs are not necessarily on their respective radar screens.

Lean Operations, Lean Thinking and Lean Construction[5]

The Lean approach for construction was first proposed by the Egan Report.[6] Nevertheless, ten years on, Egan criticised the slow progress.[7] So what is it and why does it matter? I shall not repeat what Egan said since you can readily read the original report. But what I will examine is how the Lean approach fits into construction and how this links with sustainability.

Workplaces in construction range from plush offices for developers through cool design bureaus to the somewhat less luxurious construction sites. Whatever (or wherever) it is, the workplace is where the work is done, and where value and waste are both created.

Lean construction is the part of Lean operations applied to construction activities and there are a large number of consultants out there telling you what it is. If you dig a bit deeper, then you get into the technicalities of Lean thinking. If you dig even deeper, you end up at Toyota Motors in Japan – where it was invented.

I will use the terms Lean approach, Lean improvement, Lean operations or Lean thinking in an interchangeable fashion in this book rather than Lean construction because Lean is more than just construction or manufacturing.

Lean started when Toyota wanted to be competitive against the American car companies back in the 1950s. It took them nearly 50 years to perfect this. Over those years, Toyota laid down a set of working principles that became the Toyota Production System or Lean operations as we know it. 'Doing' Lean, like 'doing sustainability', is not a one-off exercise but a way of thinking and a life-long way of doing business. The Toyota debacle of 2009 is a result of poor quality control coupled with the overly ambitious aim of becoming the world's biggest car maker and compounded by appalling public relations. The actual

5 The Lean approach is a way of working, whether it is building, managing, designing or planning. I prefer Lean operations because that is what it is. Phrases like 'Lean construction' seem to imply it is the construction component of Lean and if you are not actually building, then it does not apply to you.

6 Also see Chapter 18, Egan was a car man (he used to run Jaguar) so an endorsement for Lean was not unexpected. As for reading the report itself, it is probably easiest to type 'Egan Report Rethinking Construction' into your browser search engine to find it than for me to write a web link with the change in the UK government and its various websites.

7 Ten years on, Sir John Egan awarded UK construction four out of ten marks, as reported by *Building Magazine*, 21 May 2008.

production process is in itself still sound, even though the management control was not.

The Toyota Production System spread across Japan and soon there were the 'Canon Production System', the 'Honda Way', and later Western car companies have their own versions, such as the 'Vauxhall Production System'. The term 'Lean' came from US research on Toyota car manufacturing in the late 1980s.[8] This stuck a chord in the West and became the generic term for derivations and developments based on the original Toyota Production System.

For construction, we need to examine whether the Lean model is directly applicable. The short answer is 'yes – but with adaptations'. Construction in the UK has some attributes that are not encountered in Japanese car factories: our turnover of staff from subcontracting means our teams on site are often changed completely during every project; and every project is itself both similar and yet dissimilar to every other project. These differences mean that we need to adapt the Toyota Lean approach to suit the needs of both construction and sustainability.

What is Lean Thinking?

At its simplest, Lean thinking is about adding value, about waste and reducing it, and holds dear the principle that any effort that does not lead to value is waste.

Although Toyota did not plan it that way, there is a lot in common between sustainability and Lean thinking. Sustainability is about using our resources effectively: money; people; and materials. Lean is about reducing waste in everything we do – reducing the waste in money; the waste in people's efforts; and reducing materials waste. Lean is also about smoothly running operations – flowing effectively while optimising resource usage. You can see the connections – at a very simple level, practising Lean is more or less the same as practising sustainability and even at a more detailed level, it is still valid. Even though it is not designed to cover society's needs[9] as it was designed by a

8 The report was published as a book in 1990, *The Machine That Changed the World*, by Womack, Jones and Roos.

9 There is always a debate between the hard-nosed executives and socially progressive people about what is the role of a company. I think one of the more interesting answer to this came from the then finance director of Berkeley, the house builder. Apparently the way to keep margins high was to refuse to build affordable housing since 'that's the government's problem'

company for its internal use, the Lean approach to the workforce can be readily adapted to support the external social elements of sustainability. Both Lean thinking and sustainability share some basic drivers (see Figure 9.3).

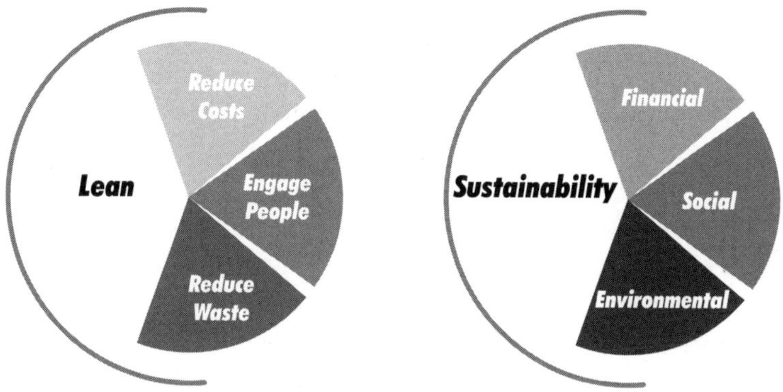

Figure 9.3 Lean and sustainability: a similar approach

There are usually four distinct drivers attributed to Lean operations: top-level support; respect for people; attention to processes and procedures; and continuous improvement. These are equally the drivers for sustainability.

Table 9.1 Drivers for Lean and sustainability

Drivers	Lean Operations	Sustainability
Top-level support	Senior champion sets the agenda and delivers the resources	Senior champion provides resources to delivers the sustainability commitment
Respect for people	Engaging everyone at work to take part	Sustainability needs everyone to make it work
Processes and procedures	Looking at the way we work to reduce waste and improve value	Everything we do is re-evaluated and the more sustainable approaches adopted
Continuous improvement	Always looking to do better	Always looking to do better

and not Berkeley's – as reported in the *Financial Times*, 5 December 2008. This man became the overall boss of Berkeley in 2009.

Where the two differ is in some of the external drivers. Lean operations is not driven by legislation although the public and governments do seem to appreciate a less wasteful company. Sustainability is driven by legislation and public sentiment. Lean comes from a competitive commercial environment where unsatisfied customers can mean the business going under. Therefore its focus will appear more commercial than the 'save the world' attitude in some areas of sustainability. Digging a bit deeper, however, the interests merge again – both sustainability and Lean operations have the same aim for the company – to ensure that the business keeps going.

The outcomes for Lean operations are also consistent with the requirements for sustainability: better use of resources. Although Lean thinking does not go so far in suggesting that any unused resources are made available for future generations, it does encourage waste reduction at the point of use. Philosophically, Lean is not about future generations or society as a whole, but instead it is about the current state of business and how to do better while sustainability is social–political movement to safeguard the world for future generations.

However, in a recession or a sluggish recovery, using Lean operations as a stepping stone towards a sustainability transformation is valid and effective. The Lean tools are generally well known, in the public domain and most importantly, well tested with employees of all types and inclinations. When applied effectively they will produce the results desirable for a sustainability transformation.

Although the financial and environmental aspects of sustainability are most readily associated with gains from Lean operation, the social elements of sustainability can also be delivered by applying the four Lean drivers towards external needs.

Table 9.2 The social elements of sustainability

Lean Drivers	In Context for the Social Elements of Sustainability
Top-level support	Champion and stakeholders respond to social trends
Respect for people	Respect the need to share the environment with the neighbours and with society as a whole and understanding the responsibilities the company has towards society
Processes and procedures	Review and revise practices that supports the accountabilities and responsibilities towards society
Continuous improvement	You can always do more/do better

The 64 Million Dollar Question: Does Lean Work?

This simple question deserves a very careful answer. First, many Japanese companies managed to get Lean to work, whether based in Japan or in the West. Then there is the experience of many Western companies who found that Lean did not really work for them since they still went bust (for example, General Motors). I know of many hospital teams in the UK who seemed to find the promises of Lean more a mirage than reality, conversely, I have also trained health sector managers to deliver Lean improvements successfully in their areas. Closer to home, Lean construction has been around since Egan's Rethinking Construction Report of 1998, what is the progress in our industry? The answer matters because if we are to base doing sustainability on a Lean foundation, we need to be sure that Lean does work.

The simple answer is that Lean does work, but it has to have serious top-level commitment and a willingness to do the hard work to make it happen. It is not necessarily easy and the conservative nature of our industry makes this harder than it needs to be. Other difficulties can include the ever-changing workforce because of subcontracting as well as the fragmented style of corporate management where the individual construction projects are essentially the construction director's fief (see Chapter 18 for more details). None of these are impossible to tackle as Lean leaders, such as Shepherd Construction and Thomas Vale, have shown again and again.

Six Sigma Quality – Hype or the Real Thing?

Six Sigma Quality has an evocative name, at once mysterious, high tech and serious. Since its invention in the late 1980s, it has been hyped to a ridiculous level by the activities of major companies, such as General Electric and Allied Signal. In the early 1990s, Six Sigma consultants were earning over £1,000 per day, which is pretty good living for engineering and operations consultants even now. Alas, all too soon, people started realising that Six Sigma is not that easy to implement and just because you have Six Sigma, it does not guarantee corporate success. Motorola, who invented it, still lurches from crisis to crisis. Polaroid, the instant film company (remember them?), another Six Sigma pioneer, went bankrupt in 2001. So what is Six Sigma and more importantly, what is the point of Six Sigma in sustainability for construction?

Six Sigma Quality[10] was invented by engineers at Motorola to reduce defects in multi-stage, multi-component manufacturing. Then consultants took over and suddenly Six Sigma (the word quality was quietly dropped) became 'the breakthrough management strategy revolutionising the world's top corporations'.[11] It is no longer just about quality improvement but also about 'making money'. Hype aside, the Six Sigma approach has some very useful tools that can contribute a great deal in 'doing' sustainability.

Six Sigma matters because it promoted a very important principle, one also valid in managing sustainability. This is about making decisions based on data and facts, or more prosaically, 'if you cannot measure it, then you cannot manage it'.

Of course, the above description is an old saying and you have heard it before, but you will pay a lot more attention if you are paying £££ for a consultant to say it back to you! Whereas there are similar sayings in Lean thinking, Six Sigma, by virtue of being a 'breakthrough management strategy', reached the boardroom a lot quicker than Lean thinking. And there is nothing wrong with senior management being reminded that decision making should always include hard facts.

Why is This so Important for Sustainability?

Much of sustainability is about dealing with problems, waste problems for example. It is all too easy for a manager to sit in an office and decide what the solution should be, since it is simpler than going down to where the waste is happening (often a much less pleasant place than the office) and finding out. Much better to 'guesstimate' what the problem is and then offer a solution. This is how things have been done for years in all areas of business. However, in sustainability, the data-centric approach is normal.

This is because much of 'doing' sustainability currently is about auditing, certification and reporting. Accurate information based on facts and data is therefore crucial and people are used to this approach. Also, since sustainability is an emotive subject and readily available for hijacking for special interest

10 The term 'Six Sigma' is derived from statistics and describes 3.4 defects per 1 million opportunities. In other words, very high quality.

11 I did not make this up, it is actually the name of a book, *Six Sigma: The Breakthrough Management Strategy Revolutionizing the World's Top Corporations* by M. Harry and R. Schroeder, 2000.

groups; having hard facts and making decisions based on accurate data are critical aspects for clear-headed decision making. (Although sometimes the data can stir up emotions even more.) Finally, as there are still many sceptics out there, the enthusiast sustainability advocate will find life easier armed with data and numbers.[12]

The Six Sigma process is famous for a lot of heavy duty number crunching since it is really a statistical process management tool. It starts to lose the 'management strategy' bits as it goes into the realm of statistics. I also diverge from this aspect of Six Sigma and move towards the people management side of Six Sigma. As one Six Sigma Black Belt[13] told me, 'All too often, the people side of Six Sigma is forgotten, it is equally important.' This is the part of Six Sigma I will focus on.

The Six Sigma approach links up well with doing sustainability as shown in the table below.[14]

Table 9.3 Six Sigma and sustainability

Activities	Six Sigma (People Aspects)	Sustainability
Problem focused:	Define problem	Identify issue
Data-backed decisions:	Gather information (causes) and analyse	Gather information (legislation, impacts) and analyse
Reduce defects/impacts:	Improve by implementing and controlling the solution	Develop solution and monitor progress
Improve quality:	Continuous improvement	Sustainability never stops

Furthermore, a Six Sigma focus on facts and real data should be adopted by policy makers everywhere because, as Trudy Harpham says, 'they need to make policy based on facts!'.

12 A rather charming 2008 IBM TV advert about green computing illustrates this: a skeptical boss turning mellow when the young bright-eyed manager blurted out how much money they save by going green, 'the IBM way'. See it at (www.youtube.com/watch?v=VSNFE6eUjfY).

13 For some reason, Six Sigma Quality uses karate rankings to describe the Six Sigma team members. I have not been able to discover the origins of this (perhaps it is just as well). The black belt has a senior Six Sigma consultant role.

14 There is one major difference in the philosophy between sustainability and Six Sigma. The latter has an inherent limit – 99.99966 per cent quality. Sustainability, however, is infinite – you can always do more and do better.

Both Lean and Six Sigma are compatible with each other to the extent that 'Lean Sigma' also exists as a combined 'breakthrough management strategy'. (Of course, this book is not so crass to claim that it is the source for another new *breakthrough strategic management* approach called 'Sustainable Lean Sigma'.)

Roadmap for Going Forward

Both Six Sigma and Lean operations emphasise the reduction of waste through process improvements and they form the basis of a simple roadmap for 'doing' sustainability.

There are four stages on this roadmap.

1. **A**gree a corporate agenda.
2. **A**nalyse the current business processes.
3. Take **A**ction to improve the existing business process and start developing a competitive edge.
4. **A**dvance towards transforming the organisation as a whole towards more sustainable operations.

The roadmap is explored in detail in Part Four of this book.

PART 3
Sustainability Policy

10

Policy and Agenda

During an early period of writing this book, UK construction industry 'welcomed' several major policy events and countless smaller ones. This section reviews some of these major events and then laments, muses or pokes fun at them as appropriate.

The Major Policy events[1] reviewed are:

- the Code for Sustainable Homes;
- the Site Waste Management Plan Regulations; and
- the Strategy for Sustainable Construction.

The Code for Sustainable Homes

This replaces the EcoHomes scheme for domestic dwellings and has six levels of attainment. One thought immediately comes to my mind, if climate change really goes the way of global warming, then is a better insulated home the way to go? There is still an attitude in the UK that it is a country that is bitterly cold,[2] despite the media telling us that virtually every year is the 'hottest year since records began'. Never mind, the Code for Sustainable Homes (the Code) has stirred a whole range of responses, from the costs of achieving the different levels, to the suitability of having mechanical ventilation.

---

1 This is my major policy event list, if your favourite item is not in here, sorry – I cannot fit everything in.

2 It is not just the government that thinks this way, I saw a TV advertisement in June 2009 from a famous double glazing company extolling their products' ability to keeping the cold out.

Basically, the Code sets out the requirements (grossly simplified here) as shown below.[3]

Level 1 homes will be 10 per cent more energy efficient and 20 per cent more water efficient than most current homes. It may also have some of the other features in the Code, such as providing office work space with communication links within the home, secure cycle storage or greater security features.

Level 3 homes would be 25 per cent more energy efficient and have many more sustainable features than a Level 1 home.

Level 6 homes would be highly sustainable and over the course of the year their net carbon emissions would be zero. Needing over 90 per cent of the points available, a Level 6 home would include most of the sustainability features in the Code.

The Code works, in principle, as follows (this is a very simplified version!):

- You design and construct a building to a particular level of Code compliance (there are 6 levels, but for most practical purposes, it is level 3–6 that developers focus on since level 3 is considered as 'current' best practice).[4]
- Get it assessed at the design and post construction stage independently.
- Once it is done, display a Certificate of Performance towards the Code for Sustainable Homes for the completed dwelling.

The reality is somewhat more complex.

In 2004, the Sustainable Buildings Task Force reported that, 'The built environment is at the heart of our economy ... But ... the way we use natural resources for building and the levels of pollutants emitted in the process of building, and in the use of buildings once occupied, are unsustainable.'[5]

3 This is from the UK Department for Communities and Local Government.

4 This is from a mock-up of a Final Certificate from the UK Department for Communities and Local Government.

5 This is from the UK Department for Communities and Local Government.

The Housing White Paper of July 2007 listed the following goals:

- more homes
- more social housing
- more affordable homes, and
- greener homes.

These goals led to thinking about developing affordable housing as well as more sustainable housing.

The Code is the single national standard for sustainable design and construction of new homes. It came into effect in April 2007 and is voluntary unless building on public land. The government implemented mandatory ratings against the Code for all new build homes from 1 May 2008. This means that sellers of new homes will have to provide buyers with either a Code certificate of assessment or a nil-rated certificate of non-assessment. The certificates form part of the now to be phased out Home Information Pack.

In 2007, *Building Magazine* (BM) reported that there were mixed views about the Code (BM, issue 25, 2007). While the government claim strong support for the Code in its press release, a quick read of the report on the consultation tells a more worrying story:[6] many questions elicited very significant 'don't know' responses. For example,

> A 48 per cent 'don't know' response to *'We believe that a zero carbon target is the most robust framework for reducing the carbon footprint of new development. Do you agree that our definition of zero carbon in paragraph 2.33 is the right approach?'*
>
> And a 51 per cent 'don't know or blank' response to *'Where there are circumstances in which the additionality of offsetting measures outside the development can be demonstrated and are more cost-effective (e.g. on small infill developments), is there a case for carbon neutrality (i.e. taking account of offsetting measures)?'*

6 Building a Greener Future: Towards Zero Carbon Development – Analysis Report of Consultation Responses, Department for Communities and Local Government, 2007.

I can sympathise with both the 48 per cent and the 51 per cent of respondents. I would probably answer 'don't know or blank' to the second example above as well.

Having looked through the report I noticed there seemed to be a lot of 'don't know' responses – a quick check showed that the 'don't know' responses were the majority 17 times and the 'yes' or 'agree' responses were the majority 6 times. I am not sure what this means to the people running the survey, but I feel that people in our sector either do not feel confident about their knowledge about the Codes or the questions are too obscure. It does make you wonder about our policy makers sometimes.

The Codes are based on a credit system and have the following categories:[7]

1. Energy and Carbon Dioxide Emissions (max. 15 credits – mandatory).

2. Indoor Water Use (max. 5 credits – mandatory).

3. Environmental Impacts of Materials (max. 15 credits – mandatory).

4. Management of Surface Waster Run-off in Developments (max. 2 credits – mandatory).

5. Storage of Non-Recyclable Waste and Recyclable Household Waste (max. 4 credits – mandatory).

6. Pollution – Global Warming Potential of Insulants (max. credit 1, non-mandatory).

7. Health and Well-being – Daylighting (max. credit 3, non-mandatory).

8. Management – Home User Guide (max. credit 3, non-mandatory).

9. Ecology – Ecological Value of Site (max. credit 1, non-mandatory).

7 Code for Sustainable Homes Technical Guide, May 2009, Department for Communities and Local Government.

These credits are then converted into percentage points through an extensive weightings system.[8]

To summarise, the big hitters in the rating systems are energy and water use as shown below:

Table 10.1 Energy, water and the rating systems

Code Level	Energy (% better than Part L, 2006)*	Water (l/person/day)	Other Points required
1	10%	120	33.3
2	18%	120	43.0
3	25%	105	46.7
4	44%	105	54.1
5	100%	80	60.1
6	zero carbon	80	64.9

* Part L of the Building Regulations 2006 is about energy efficiency.

Source: Code for Sustainable Homes Technical Guide, version 2, Department for Communities and Local Government (DCLG), 2009.

The energy efficiency timetable are:

- level 3 by 2010
- level 4 by 2013, and
- level 6 by 2016.

Non-domestic buildings, although not in the Code, are expected to follow suit with an aim of all new non-domestic buildings to be zero carbon by 2019.

Now that we have covered the basic points about the Code, it is time to turn to some issues.

8 The weightings, according to DCLG, 'have been derived from extensive studies involving a wide range of stakeholders who were asked to rank (in order of importance) a range of environmental impacts) ... Stakeholders included international experts and industry representatives'. The Code for Sustainable Homes, setting the standard in sustainability for new home, DCLG, February 2008.

Costs of Compliance

DCLG recommends the following cost estimates for an end of terrace house for 2008 and 2016.[9]

Table 10.2 DCLG cost estimates

	2008		2016	
Level	best case	worst case	best case	worst case
3	£4,927	£5,801	£3,934	£4,987
6	£3,1207	£37,668	£2,2475	£28,293

Now these are likely to be VAT-free builders prices and home buyers are going to pay somewhat more for this: the VAT plus any margin. *Building Magazine* reported (issue 22, 2009) that level 4 has a premium of 20 per cent over base costs and for level 6, a 35 per cent premium. This hit on the wallet makes it harder to consider especially when the country is in a recession. (Of course, by 2016, things will be different, but we still have to get through the depressed economy first.)

Defining Zero Carbon

The 2008–09 recession has resulted in severe retrenchments in the house building sector. The building sector was concerned that meeting the 2016 Code level 6 targets may not be possible with purely on-site generated renewable energy (BM, issue 28, 2008). In December 2008, the UK government appeared to dilute the definition of zero carbon to include off-site generated renewable energy. Will this open the door to bizarre offsetting schemes where solar panel arrays in the Sahara will provide electricity for the UK via a giant cable?[10]

Shifting the definition of zero carbon is causing some angst, but in my honest opinion(!), I believe that once the economy settled down and growth resumed, any slackening will be tightened up. Also technology is on our side,

9 Cost Analysis of the Code for Sustainable Homes, Final Report, DCLG, July 2008.
10 Apparently the solar panel array, about the size of Wales, was dreamed up by the European Commission – as reported by the Guardian newspaper on 23 July 2008.

and there will be newer, cheaper technologies to lower the costs. Of course, whether these will be produced in China and shipped to the UK is another matter.

Other than the cost issue, there is also a vocal group in our industry suggesting that the Passivhaus, a German energy efficiency standard, may be a better way to zero carbon.[11] Basically, the Passivhaus standard reduces energy use through a very well-insulated, air tight building. What is more, it does not require expensive renewable technologies. However, since the Passivhaus is only an energy standard, it is unlikely to replace the Code in its entirety. Besides, the Passivhaus will not even meet the Code's level 3 despite being super insulated – apparently because the UK measures energy performance differently to the Germans. I think what it means is that opponents to the Code have another stone to throw at the government.

One thing that intrigued me is the need for a pressure differential between the inside and outside of the house to enable effective heat exchange. This then requires a fan or some sort of mechanical ventilation. And mechanical ventilation requires filters and that is where the next issue arises.

In late July 2009, the UK government redefined zero carbon again. I gave up at this point and decided to worry about it as and when I need to use it for any purpose.

Ventilation and Filtration

In May 2009, *Building Magazine* (issue 21, 2009) reported that the UK government is considering new building regulations after reports that ventilation systems required by Code level 4 upwards may cause health concerns. Apparently there is a problem in that people are not replacing the filters in these mechanical ventilation systems. This can lead to a breakdown of the ventilation scheme and result in a building up of humidity, carbon dioxide as well as increasing energy use. It is also possible that clogged filters will lead to breathing problems.

What does all this mean? Well, new builds will be at a very high sustainability specification by 2016 or sometime beyond. The rest of us in existing dwellings (the majority of homes) will either dread any home improvements (where we need to meet current specifications) or consider moving to new homes.

11 You can find out more by searching for Passivhaus on your internet search engine.

Of course, there will be difficulty in learning how to use the new equipment. Apparently there are credits to score in the Code for a Home User Guide – this is good news as Mindy Hadi from BRE said, 'people just don't have the necessary information on how to use some of these low carbon technologies'.[12]

For the moment, Mike Malina of Energy Solution Associates, a consultancy, mused that the current economic climate makes Code level 4 and above just a bit too expensive. Although he sees further improvements in energy efficiency, he was concerned that some of the technologies may not be ideally suited for the intended applications even though they scored the right number of points in the Code. This view was also echoed by Thiru Moolan and Rhian Williams from the Wandsworth Borough Planning team. Neil Johnson, when he was at Taylor Woodrow Construction, saw a different problem: the lack of certainties in housing development does not work well with the Codes schedule. Since the entire house building process can take a long time, then something planned for Level 4 of the Codes may by the end need to comply with Code 5 level. This type of uncertainty meant that cost planning gets much harder.

Glen Irwin, formerly director of Foreman Roberts, a building services consultancy, felt that sometimes politics are mudding the picture. He told me that we are not getting all the efficiencies that we can. For example, the planners at the Greater London Authority requiring new build offices to have 20 per cent of their emissions offset by using renewable energy. However, this meant that despite the best intentions, a lot of London buildings are deploying only renewable technologies, which may not offer the best carbon savings overall. Irwin was confident that good energy efficient designs can reduce emissions by nearly 40 per cent without using any renewable energy sources, but as these designs do not contain the required 20 per cent renewables, they may not receive planning approval, even though they nearly double effective CO_2 reductions overall. Andrew Mellor of PRP, an architectural practice, felt that some consistency in applying the Merton Rule[13] would be helpful to make doing sustainability easier.

Thiru Moolan and Rhian Williams told me a few years ago that many fashionable renewable energy equipment, such as urban wind generators, were just not at the stage of technical development where they can really make

12 You can read more about this in *Constructing the Future*, issue 40, summer 2009, BRE.

13 The London Borough of Merton pioneered the requirement of 10 per cent energy generation from renewable sources. This landmark decision has since led to other local authorities competing to 'do better'.

a difference. Williams said that, for example, wood chip biomass boilers are not necessarily the best way forward for urban areas like London even though they are the least expensive option to fulfil the renewable energy pledge. The biomass feedstock has to be transported in and there is also the need for a backup gas boiler. (This view was echoed by many M&E specialists who say there are better energy efficiency solutions than burning wood chips.)[14]

The new homes will have better energy efficiency technologies but as Mindy Hadi found out, having newer technologies do not mean people can use them effectively. Andrew Mellor saw the education of tenants as key for achieving sustainable development. This meant not just knowing how to 'operate' the home effectively but also understanding what the energy and environmental assessment of the home actually means. The Wandsworth Planners shared this view. Although sustainability can also benefit from relevant use of Section 106 schemes, Moolan felt that the community needed to have more awareness of sustainability issues, such that they can question their elected officials on how their local authority deals with sustainability. The greater awareness will also help individuals to make up their mind on investments, such as solar water heating as a cost-effective solution.

Overall, the Code is a good thing except I have one thought: the Code is encouraging (or rather mandating) the building of sustainable homes, but not stipulating whether the building methods should be sustainable or remain biblical. I am not sure how building methods can be added to the Codes, but nonetheless, in this, an opportunity is missed.

The Site Waste Management Plans Regulations (SWMP)[15]

The Regulations were published in February 2008. The SWMP was designed to stop (or limit, depends on whether you are an optimist or a pessimist) fly-tipping.[16] This is a major issue in the UK as piles of construction waste all over the countryside (alongside abandoned fridges and sofas) are not desirable as it deprives the relevant bodies the opportunity to collect taxes and other fees for

14 Sometimes, political correctness can be overwhelmed by reality; the plans to make the London 2012 Games the greenest included a wind generator in Stratford. There is only one problem: there is not enough wind in Stratford to make it work. Oops.

15 The Site Waste Management Plans Regulations can be downloaded from the Office of Public Sector Information website, or by typing site waste management plans into your search engine.

16 I recall hearing about this at various workshops on SWMP in 2006–07.

disposal services (among other more high-minded reasons). But by the time the Regulations came out, small projects were exempt and the focus appeared to be materials management.

Various industrial bodies, including WRAP[17] and BRE, developed a range of useful tools to make your day easier when dealing with SWMP and the government published guidebooks and contracted consultants to run workshops to explain what was needed.

In February 2008, the legislation was published and suddenly there is an uproar. Although it is expected that the work involved in complying with the SWMP will be onerous and immensely time consuming, the exact level of onerousness and time needed was a surprise to some. My reading of the legislation led me to think that this is not really going after small-scale fly-tippers but more like having a go at the foundations of capitalism instead!

The mandated requirement of declaring and reporting the expected waste from a construction project provides the other interesting point about the Regulations. The client and principal contractor are expected to estimate the waste levels of the project. Philosophically, this is a can of worms. Imagine a contractor telling the client that its process has a waste level of X per cent with an implied comment that the client is paying for waste (I mentioned this earlier in Chapter 8). Although we probably never think about it, a certain percentage of every product or service we purchase is waste. You can look at it like the Emperor's new clothes – it is completely acceptable until it is pointed out, then it becomes rather less palatable.[18]

If I am a client, would I then ask the main contractor why I am expected to pay the X per cent for waste? Or do I just live with the fact and get stressed? Luckily, the reality is that our industry is pragmatic and not bothered with this type of philosophical issue ('we have a schedule to meet'). So it is likely that the spreadsheets will be filled in and bland comments added to explain whatever discrepancy between the estimate and actual figures. The cost is swallowed or passed on, and life goes on, albeit with a lot more data needing storage space.

However, I wonder what the client's shareholders will make of this – publicly available records of their company paying for waste. With the UK

17 WRAP is the Waste and Resources Action Programme, funded by the government and partly at one point, I believe, by Landfill Tax money.

18 If you do not know this parable, look it up on the internet.

government bringing in its round of public sector cuts in spending, what would the government's shareholder say to this? Cuts in services because we cannot afford it? Alright. But still paying for waste during a cut in public services? Whoever drafted these Regulations perhaps should have consulted beyond their own circles.

Projects below £300,000 are exempt from SWMP compliance, so that takes out most of the small-scale refurbishment which many in the industry consider as the source of much fly-tipping. This somewhat detracts from the original goal of the SWMP, but let us not worry about this now. The real issue is the need for main contractors to estimate the expected waste before the project starts and also to report on any deviation from these estimates, and giving reasons for these deviations (see Figure 10.1).

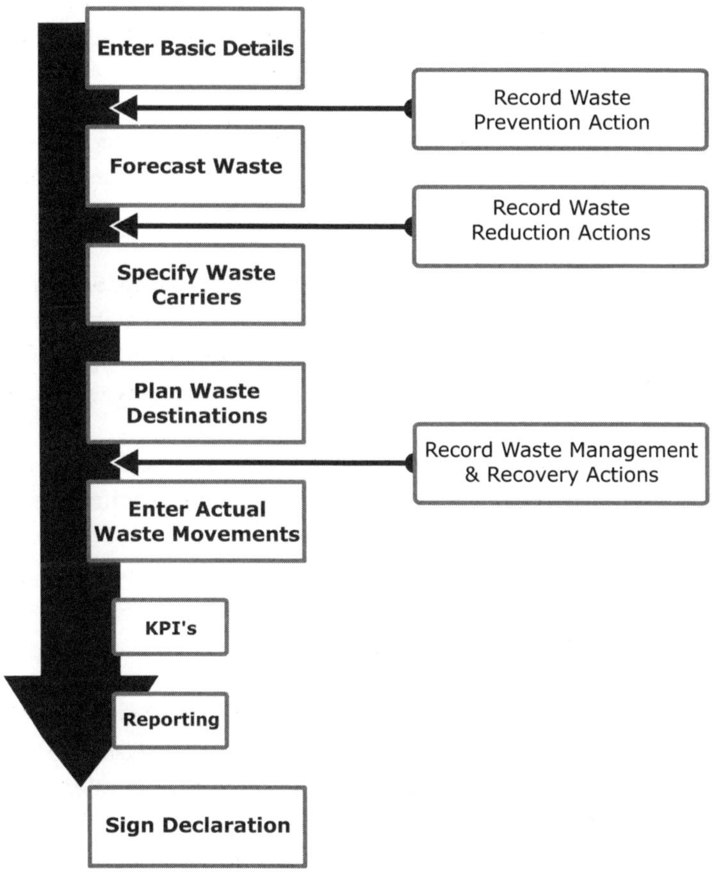

Figure 10.1 Site Waste Management Plan procedure

The SWMP makes it a legal duty for the client to report on the management of waste, and if a project of greater than £300,000 (ex VAT) is started without a SWMP, both the client and the principal contractor (if there is one) are both guilty of an offence.

A Whirlwind Tour of the SWMP

The Regulations' requirements are:

- Identify the people responsible (client, principal contractor and the person putting together the plan).
- Before drafting the plan, record the decisions aimed at reducing waste during the project (design, construction methods and materials chosen). There will also be a declaration of an intent to minimise waste.
- Then comes the plan: details about the type and quantities of all expected wastes together with the proposed waste management techniques.
- The plan needs to be updated with the usual waste transfer notices as well as paperwork on the waste management/handling operator; regular progress reports (minimally every six months) with details on amounts and types of waste, and how they were managed.
- When the project is complete, the plan needs to be updated again with explanations of any deviation from the original plan as well as data on money saved (if any).
- The plan is made available to authorised persons for up to two years after the completion of the project.

There are SWMP variations for projects costing between £3–500,000 (less bureaucratic) and for projects costing more than £500,000 (more onerous). There are also the dire warnings of non-compliance ('guilty of an offence'), the penalties (a fine up to £50,000 on conviction or fixed penalty notices of £300) and enforcement.

Clearly like the simple description of the Code, the reality of the SWMP is somewhat more complex. WRAP offers free templates on how to manage SWMP and BRE offers the free online SMARTWASTE package.[19]

It is quite clear that compliance will be generally quite laborious if not onerous. The WRAP spreadsheet suggests the following steps during the project:

1. Enter basic details (waste stream and material types, etc.).
2. Forecast the amounts of waste.
3. Declare the waste carrier(s).
4. Plan waste destination(s).
5. Record details of actual waste movements.
6. Sign the declaration (remember, not doing so is an OFFENSE).

The Plan requires reporting of the cost savings from completing and implementing the Plan (paragraph 8.4.d) as well as explain any deviation from the Plan (paragraph 8.4.c).

This is taken by WRAP as recording the cost of disposal (estimated and actual) from the waste carriers. BRE prepared a cost estimate in 2007[20] when they established that it will cost about £760 to carry out a SWMP for a £250,000 project and £1,600 for a £1.6 M project. How these numbers will work for a multi-million pound project is less easy to judge.

SWMP is onerous but what is the full corporate cost of actually implementing it? The full cost of implementation is not really a problem for the government since it is a problem for the client. The client will either try to reduce the cost of compliance from its principal contractor or pass it on to the end-user of the building. Either way, someone gets hit in the pocket.

19 Search for 'WRAP SWMP template' and 'BRE SMARTWASTE' respectively through your search engine.

20 Reported in *Construction News*, 17 May 2007.

I am not sure how SWMP will help a company trying to add real value or utility to the client. SWMP compliance can be seen as non-value-added work since it does not actually add to the utility of the building. It may enable the building activities to be in compliance with the SWMP, but it does not make it more energy efficient or more comfortable. In Lean operations terms, such non-value-adding work is a type of waste.

Going from the sublime to the somewhat ridiculous, I had a quick check about the actual wording in the Regulations:

- Incidences of the word 'reduce': zero!
- Incidences of the word 'minimise': 1 (the nearest stand-in to 'reduce' in this document).
- Incidences of the word 'reuse': 2.
- Incidences of the word 'recycle' or 'recycled': 2.
- Incidences of the word 'offence': 21.

I guess since it is the law, the word 'offence' has to be there, still it seems somewhat disappointing against zero (or one) mention of reducing (or minimising) waste. If you are a conspiracy theorist, you may think there is something going on here. Is the 'sustainability sector' against reducing waste? Has the waste management industry being doing a lot of marketing? Or is it because the government wants its share from the Landfill Tax?

To continue with my idiosyncratic scrutiny of the Regulations:

In paragraph 6.4.c, the details of waste management action for each type of waste are revealed: 'reuse, recycle, recovery and disposal', but why is 'reduce' not included? I find it unfortunate that there is no need to record any plans to reduce particular waste streams.

Paragraph 6.5.b says that 'materials will be handled efficiently' – this reads a bit like auntie telling you what you must do. There is a requirement for client and principal contractor to take 'all reasonable steps' to ensure this. The fudge word is of course 'reasonable'. This get-out clause lets the government appear to be tough, but also lets contractors to define what is reasonable. In the meantime, Landfill Tax revenues and Waste Management company incomes go up.

The choice of words is unfortunate, perhaps the government is borrowing jargon from centrally planned economies. I always thought in market capitalism, efficiency is a choice. Sure, you will go out of business if you are not efficient, but surely it is not up to the government to declare this to be a necessity.

I also had a look at a WRAP guide book[21] on SWMP. These are usually fairly good and this one in particular covered all the details a construction company needs to fill in the forms. Yet ... all its minimisation guidance is about materials management – the quick wins WRAP recommend are about waste materials recovery, not better or more efficient communications and planning, or about eliminating the Hidden Construction Site.

SWMP – a Missed Opportunity?

Personally, I think the SWMP may not encourage behaviour change since onerous form filling and reporting are unlikely to leave enough time to consider behaviour change towards reducing waste. This is made more evident especially since almost all your competitors have to do it too. Efficiency benefits are unlikely to be an incentive to improve the SWMP process since the cost of implementation based on BRE's estimates are rather low.

The discomfort I have with the SWMP is that as it does not change the fundamental thinking about site waste; then what is the point of it, other than creating lots of recycling and gathering lots of data? We need to tackle waste and sustainability from a broader perspective than environmental management. I do not think ticking boxes is the best way to ensure people to bring about sustainability, it is probably an easy place to start.

Having a framework in place does not necessary bring changes. This is something that is often highlighted by voluntary schemes, such as Considerate Constructors or even Investors in People, or the ISO quality standards, or BREEAM.[22] I am always a bit sceptical about whether these frameworks really

21 This particular one is Achieving good practice Waste Minimisation and Management, WRAP – probably 2008 although no publication details is available.

22 BREEAM is the BRE Environmental Assessment Method, a commonly used set of standards for improving environmental standards in buildings. A (sort of) UK competitor to the US LEED system.

bringing about behaviour change, as opposed to being basically ticking boxes and then sticking a plaque on the wall. So it is back to putting up a sustainable building but not building in a sustainable way.

In giving evidence to the House of Commons,[23] sustainability experts Jon de Souza, of Constructing Excellence, and Andrew Kinsey, of Lend Lease, reported on compliance with the SWMP regulations. There was a hint that the monitoring bodies may be too overburdened to take much of the SWMP reports into account. My further investigations on the actual mechanics of managing the SWMP showed that in some cases, the SWMP efforts is more a demonstration of completing checklists than actually doing something about cutting the waste.

I am surprised that clients and contractors are not seeing the initial waste estimates as a challenge to reduce wastes. But I was told that clients do not often see waste as part of their business outcomes and neither do contractors, both wanted to focus on getting their prime job done. Therefore if the inspection regime is as relaxed as my sources seem to indicate, not a lot will happen other than unread reports piling up somewhere. This is a pity as the initial SWMP estimates should provide an initial target for improvement. I guess this is because of the following reasons:

- Waste is tied to environmental management, not to business efficiency.
- The SWMP process is onerous and companies get so tied up in the mechanics of it that they lose sight of the real bigger picture. (Although from reading the legislation, the bigger picture is just to report on materials movements – which is being carried out.)

Perhaps we should have a look at the construction materials flow process before we leave the SWMP.

- Construction materials usually use a lot of energy to make (think cement, glass, steel, bricks) or are reliant on a non-sustainable source (plastics), or takes a while to grow (timber).

23 This is from the 'Uncorrected Transcript of Oral Evidence to be published as HC-1100-iii' taken before the Environment, Food and Rural Affairs Sub Committee on 19 November 2008.

- We then transport a lot of this, whether globally or locally, it still means a lot of traffic, using fuel, causing congestion and pollution.
- When it arrives on site, we waste about 100 M tonnes of it every year. (Every fifth or sixth pallet load.)
- And we pay people to take it away for recycling (cause more traffic and pollution, using energy to reprocess and then to transport again) or pay for it to be hauled away and then pay tax on this.

Look at it this way, the construction materials flow process is a costly embarrassment and yes, there is a missed opportunity.

Are the lawmakers responsible for this missed opportunity? Or is it their advisors in the industry that led everyone down the path of filling in boxes? Who knows? I can only guess that costs for the industry will go up without much discernable gains in efficiency.

Strategy for Sustainable Construction (the Strategy, for Short)

Ah, the Strategy for Sustainable Construction![24] This should be the centrepiece of this book. But in June 2008, the industry's focus was somewhat diverted by the on-going credit crunch and looming recession. The industry comments since its launch has been – nearly zero. A quick search on the *Building Magazine* website has found only 17 matches for Strategy for Sustainable Construction from June 2008 to June 2009. (Surprisingly, *Construction News*' website also returned 17 matches from 11 June 2008 to June 2009!)

My aim was to read this document thoroughly and present a proper review, but it is not a light read. No, not in the usual light/ heavy sense (let's face it, what government strategy is ever light reading?). No, it is so full of acronyms that it is like reading a foreign language journal. You stop every few seconds to think 'who are these people again and why do they matter' or even worse, 'what do these three or four letters mean again?'

My mind was being overwhelmed by the range and breadth of PPS25, CRC, CERT, SDC, DCFS, CQS, NESTA, EP, CEN, DIUS, TSB and their assorted

24 You can download your very own copy from the website of the Department for Business, Innovations and Science (or whatever it is called this week).

cousins, and it became harder and harder to concentrate on the multi-colour pages. Simply because of acronym overload, I started skipping over the incomprehensible letters and, of course, then the Strategy made even less sense. The last time I felt like this was trying to understand the Stern Review. I have to say that with the current economic climate, like everybody else, I am just not paying as much attention to the Strategy than had it been in the good times.

The Strategy forms an action framework for the industry and it can also act as a checklist for companies wanting to be progressive as well as tendering for work with public sector clients. To this end, the aims and targets in the procurement, design, innovation, people and better regulations sections can probably help the contractor (well, not the better regulations sections). This group form the 'means' and the 'ends' are: climate change mitigation and adaptation; water; biodiversity; waste; and materials all can contribute to helping companies realise where the government's thinking is heading in sustainable construction.

Thoughts on the Strategy

The Strategy starts with some statistics, such as 'Buildings are responsible for almost half of our country's carbon emissions, half our water consumption about one third of landfill waste and one quarter of all raw materials used in the economy.' This is useful stuff for future presentations, but that is about it. Some of the targets, however, induce a sense of amazement:

> *'25 per cent reduction in the administrative burdens affecting the private ... sector'*. This from a government who just introduced the Site Waste Management Plans?

However, getting away from the 'means' to the 'ends' part of the strategy does bring some relief as the targets become more concrete. Like a target of 130 litres of water use per person per day by 2030,[25] or reducing total UK carbon dioxide emissions by at least 60 per cent on 1990 levels by 2050.[26] And a 50 per cent reduction of construction, demolition and excavation waste by 2012. Then there is the zero net waste. Wait! There is *no* mention of zero net waste! We

25 I assume this means most of us will not be living in Code 6 housing by 2030, since Code 6 stipulates 80 litres per day in new builds by 2016.

26 Does the average person know what this is in a real-world sense? I do not. Are we going for nuclear energy? That will really cut the CO_2 emissions, either that or cover the UK in solar panels and wind generators.

moved smoothly from the 2007 draft Strategy suggestion of zero net waste at the site level by 2015 to zero mention in the full Strategy. I guess nobody wants to be responsible for something that is really rather hard to sort out.

Then there are the responsible sourcing schemes. These are good ideas but also represent an increased in costs in assessment fees,[27] in compliance and in administration including buying extra box files to hold all the certificates.

What about existing buildings? There are no targets for these but a policy framework that comprises 'actions to inform, support and incentivise those who are prepared/able to change their behaviour and/or take actions in their own homes'.[28] I hope this means the Energy Saving Trust will in future give advice for apartments with flat roofs, which I did not get (*'we don't deal with flat roofs'*).

So here is a summary of my thoughts.

In the real world, being responsible means 'being responsible' and in the commercial world, it means you can get sued if you do not deliver. In the Strategy, responsibility is handed to organisations that do not seem to have any real accountability over the events. I cannot see how the Strategic Forum can be the body responsible for '34% of client activity, by value, embraces the principles of the Clients' Commitments'.[29] How is the Strategic Forum supposed to make sure that happens – for real? The Strategic Forum, according to its website, 'does not act as an operational body. Its strategy is being implemented via Constructing Excellence, ConstructionSkills, the Construction Umbrella Bodies, National Platform and other implementation bodies together with the wider stakeholders in the construction industry where appropriate'.[30] But none of these organisations have the authority to make anything really happen if the industry chooses to ignore them.

What will happen to the Strategic Forum if it fails to deliver these actions? Will it be penalised or lose its funding? Will its executives resign on mass?

27 I cannot see how extra certification will mean same or lower costs, unless the assessment bodies are doing it for free; which is unlikely.

28 I liked to be incentivised on this: I thought about fitting the energy efficient glass years ago, but was deterred by the cost £2,000 to cover 11 windows; properly insulating my flat roof was quoted at £7,600 (all ex VAT too).

29 Strategy for Sustainable Construction, HM Government 2008, Actions and Deliverables for Procurement.

30 Strategic Forum for Construction website (www.strategicforum.org.uk/).

Unlike Wembley Stadium, it is unlikely the government will drag the Strategic Forum to court for non-delivery. Clearly in the world of strategies, being responsible is not the same as being responsible in the real world.

Do I have a wish list for the Strategy? Yes, I would really liked to have seen the policy makers and the decision shapers (whether in agencies, trade bodies or government) take a work placement/job swap with project managers in construction projects. We can then give the policy making people a dose of what their regulations really do and the projects people an understanding of the workings of policy making. Any other wish list items? Well, there is always a wish for less red tape and a smaller government footprint.

The Strategy did not get the reception it deserved and those who responded seemed to say negative things.[31] But I also know it is not easy to prepare this document with a team of two people or however many survived the various DTI (BERR or BIS) staffing level cuts. So it is congratulations to the valiant few who put this document together because we need a set of guidelines like this Strategy to start the serious discussions. But I would still like to know how the Strategic Forum is going to really contribute to making the accident rates drop 10 per cent year on year.

Good work but very bad timing (and its not your fault, DTI (BERR or BIS), Defra, CLG, DIUS, DCMS, OGC and the Strategic Forum). Hmm ... is this a case of perhaps too many chiefs?

31 *Construction News* reported on 11 June 2008 that the opposition party (now the government) spokesmen on construction reckoned the (then Labour) government is 'inconsistent and confused'. I guess we will have to wait and see whether the UK's current coalition government is any less inconsistent and confused.

PART 4

Doing Sustainability

11

Going Forward

The previous chapters examined the background to sustainability in construction: the ways of satisfying customers and also about managing the relationships with stakeholders. I now focus on two areas of going forward towards a more effective and sustainable construction business. The two areas are improving what we are doing now and transformation – developing a new set of policies and strategies for sustainability. A generic way of going forward, is shown on Figure 11.1. The need for transformation to a future state is matched by the need for improving the current state. Before all that can start, however, we need to get the support of the senior decision makers.

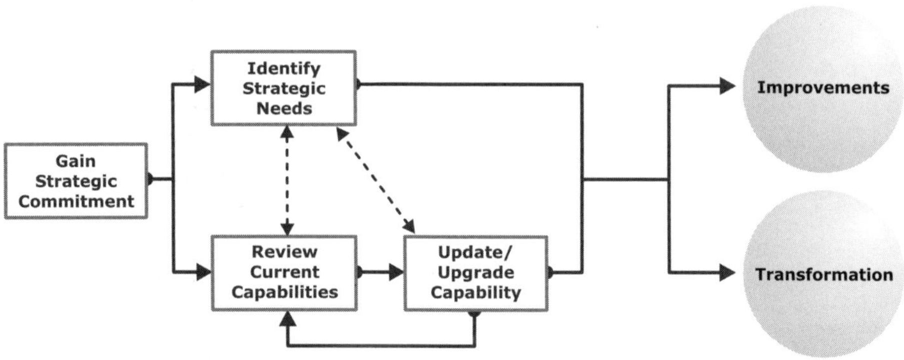

Figure 11.1 Going forward to sustainability in construction

Gaining Strategic Commitment

Despite the efforts of sustainability specialists, legislation and social pressures, the feeling that many senior managers are not too committed to sustainability persists. As I mentioned earlier, it does not really matter whether you believe it is claptrap or not, the reality is that we need to manage sustainability pragmatically and professionally. This means making sure that any personal scepticism is not showing. This is because people are perceptive and a subconscious message is sent and received. I believe an example of this was where the HR manager of a major contractor, told me that he could not fill his health and safety training courses as the construction managers claim they are too busy. When excuses like this are tolerated even though health and safety is backed by legislation and coupled with the threat of criminal liability for construction company executives should things go wrong, it seems to me that the staff has realised that the company is not really that serious about health and safety. Therefore, it is still possible to wriggle out of the necessary training.[1]

Top-level support is something that always needs to be done and done often, judging from the current support for sustainability, this appears more lip service than a realistic budget. Is this a failure by the sustainability professionals? Or is that the sustainability message, not withstanding eco disasters, legislative push and social pressure, just not compelling enough? Or is it something else? As Neil Johnson said, 'embedding sustainability is not always easy'. Currently it is still commonly seen as a 'soft and fluffy thing' and therefore difficult to grasp. Although the Codes for Sustainable Homes have concentrated the mind, Johnson believed that a company needs to see sustainability to be actually 'good' for a company – it is not just about feeling good or giving good PR. He went on to say, 'This is not easy to do as companies have an internal inertia for this type of thing and it is not helped by the overwhelming amount of information on sustainability. This huge amount of data is neither easy to use nor always consistent. The result is often a difficulty in training the workforce and sharing the understanding. People have to resort to personal contacts to cut through the data in order to gain an understanding.' It seems like we, through our own hard work, is making it harder for people to grab hold of the topic.

But there should not be any doubt about the prospect of future eco disasters if we do nothing now. There is also no doubt about the interest from our society, however overwhelmed with information. The sustainability professionals I know are all enthusiastic and knowledgeable. So what is the problem?

To me, there are several reasons why real strategic commitment is not always there. First, eco disasters are always in the future. Second, changing

1 Of course, the construction managers may know better: it is possible that those specific courses contained little value to site operations but merely allowed the HR department to tick its boxes.

is hard. Third, we do not always explain sustainability in context. Fourth, embedding sustainability into business processes is still on-going, but not pervasive yet.

1. ALWAYS TOMORROW, ALWAYS SOMEWHERE ELSE

Eco disasters generally happens somewhere else if it is happening now; or it will happen by 2050. With the UK's business culture being short termed and the construction industry focusing on the current projects (usually two years or less), how is sustainability going to impress itself on our consciousness? Our loans are maturing in a year, the Inland Revenue wants its cut in three months, salaries have to be paid in four weeks. The timescales are just not in favour of worrying about eco disasters.

Even if the waters of China are polluted by its industry making products for the UK and droughts elsewhere are forcing migrants to come and seek work on UK construction sites, it is still a long way away from us in Stratford or Leeds. We even have proverbs saying 'a bird in hand is worth two in the bush'. We are short-termed creatures! Therefore, eco disasters by 2050 will probably make us feel a bit bad, but so what? Will the Inland Revenue be sympathetic if we say our budget is diverted from paying taxes to repaying the global eco system? Investors want returns now – who wants to wait until 2050? We may be dead before then. Therefore, talk of eco disasters tend to result in lip service, but little serious budget. Even when it is Al Gore or a rock star hectoring you, you may buy a T-shirt if you like the rock star, but you saw the Al Gore movie twice, is not that enough?

Why do we say eco disaster will happen in 2050? Because had we said 2013 and we ended up with another 'phew! what a scorcher' summer or a cool wet summer, the experts will look bad by making wrong predictions; not hot enough or too hot; not wet enough; or too wet. The fact is that eco disaster is happening right now, perhaps at somewhere else in the world today, but it is definitely happening now. The problem is not the timing of the eco disasters or our short-termed vision, but probably more our inadequacy in marshalling our arguments effectively. Sometimes, those with the knowledge also seemed to want to shape our knowledge too, and when they slipped up, the rest of us become either more disillusioned or more despairing. The response from many over the climate change email scandal which broke in late 2009 is that it is an example of self-inflicting wounds by a group of elite scientists, perhaps a bit to sure of themselves.[2]

2. CHANGE IS HARD

Change is hard, even crossing our arms the opposite way[3] or changing our brand of toothpaste can be uncomfortable. Therefore asking people to change the way they work is even harder. Sustainability requires a lot of changes in the way people, processes and companies work. It is not enough to stop printing emails or segregating waste on site. It is about a wholesale commitment across the company towards change and that is hard.

Since it is hard, not everyone wants to do it. Especially since the payback is preventing an eco disaster in 2050. When not everyone wants to change, then change happens at the pace of the slowest mover – in other words, nothing much happens. Change can be made easier if there is some relevant incentive as well as the right information, which leads us to the next point.

Regulations like the Site Waste Management Plans are a good example of half-hearted change that somehow missed the point. A good intention got diluted by perhaps the slowest movers, perhaps special interests or perhaps by a simple lack of imagination and we end up with red tape and bureaucracy. Change is hard enough on its own and getting bureaucrats to lead change just makes it harder. How can we convince senior managers and clients to change their behaviour when the legislation just encourages a tick-box mentality?

3. SUSTAINABILITY IN CONTEXT

A lot of my friends cannot understand how I can be writing a book on sustainability: I eat meat! I like cars! I go on overseas holidays! I like incandescent lights and so on. How can someone like me be a cheerleader for sustainability?[4] It should have been someone who has perhaps a bit more formal knowledge of environmental management, someone who prefers bicycles to cars, someone who wears organic fabrics, someone, well, someone more like themselves.

---

2 There were press reports from November 2009 to early 2010 on leaked emails, disguising inconvenient report findings as well as simpler mistakes like getting some of the facts wrong.

3 See Figure 15.13 in Chapter 15, for more on this.

4 It gets worse, I was taken to task by a colleague for not recycling my worn out underwear. 'Nuff said!'

To me, that is one of the difficulties of promoting or marketing sustainability, whether cajoling or admonishing, the message is not really well tuned. The cajoling: it is good for whales, trees, seals, aquifers and you of course; it saves you money; it will save your kids; it is socially acceptable – all that becomes noise after a while. The admonishing: you will be fined; sued; spat on in the street; die from food shortages; die from new diseases; killed in heatwaves; killed in floods; you will cause the extinction of insects, fishes, animals, people and society ... This also becomes part of the ceaseless marketing we get bombarded with every day. The voices telling us are often self-righteous and sometimes intolerant. Unfortunately, these messages are far from effective because they are not tuned to our personal needs. Similarly, our senior managers hear this stuff and they respond by making some polite noises and then tune it out to deal with more immediate issues, such as delivering a building on time, to cost and to specification.

As Neil Johnson suggested, 'we are inconsistent and we produce so much data that no one has the time to sift through and find what they need'. Katerina Maluenda and David Hopwood of Kew Design/Build, an architectural practice, told me that a small practice like theirs finds it very difficult to find the right information simply because there is so much unsorted information out there.

Sustainability is about as universal as buildings, but every building is aimed at some individual client's need, why can we not differentiate our sustainability message the same way?[5] The 'same' is just boring, most people who can afford it will want to avoid living in identical pods, so we decorate the insides and outsides of our buildings differently to allow for individual taste. So why should the sustainability message be the same for everyone? I have a somewhat controversial assumption on why we use mass broadcasting when sustainability needs the opposite – a tailored delivery. It is, I believe, because the current generation of sustainability professionals are made up of specialists. The group of people making decisions on sustainability are often trained in environmental management or are self-taught and shaped by environmental legislation. To me, this monoculture results in a lack of imagination, poor marketing and not able to firing up people's enthusiasm generally.[6]

---

5 I know, the Rethinking Construction Report by Sir John Egan (see Chapter 18) suggested more common designs, but let's take a reality check: the style gurus will never allow it!

6 Of course, many colleagues do it very well. But to be honest, sustainability has never been well marketed. Energy efficiency which started around 1973 and is still not everywhere now as compared to mobile phones and internet deployment, which are both much younger but everywhere in the UK. I put it down to more effective promotions and targeting by marketing professionals and not being told what to do by technologists.

Luckily, sustainability is part of everyone's life and by the next generation, everyone should have been trained in their professions with sustainability designed in. For the moment, the sustainability professionals need to embrace broader marketing tools to get their message across. They also need to embed their message into the broader streams of business operations and to do that, the sustainability professionals need to be in the business, not in corporate sustainability silos.

4. NOT SUFFICIENTLY EMBEDDED INTO BUSINESS OPERATIONS

If we have sceptical senior executives because eco disasters seem a long way off, because the thought of changing from 'business as before' to 'business as it should be' sounds a little too tough, maybe because we hector them a bit too much and our messages come on a bit too strong, it may be time to have a rethink. I advocate a rethink about our tactics only, not our beliefs since we are, of course, right (see Figure 11.2).

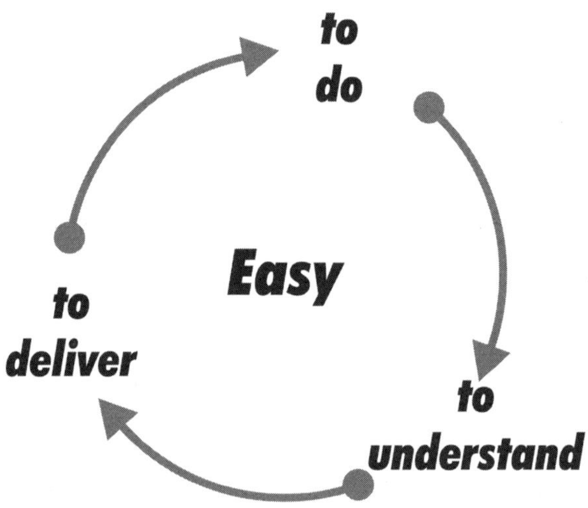

Figure 11.2 Making changes as easy as possible

If the obstacles in front of us seem a bit daunting, maybe we should try to go round them. I suggest we make changes as easy as possible to get buy-in. Easy comes in several self-reinforcing flavours:

Easy to do: it is not rocket science and does not require a lot of hard changes or if there are tough changes, then the benefits should be easy to understand.

Easy to understand: the message is in context and so connects with the people you want to change. If the message cannot be readily customised, then getting good results need to be easy.

Easy to deliver the goods: there must be easy targets that everyone can share in the success. If the results take a while to arrive, then it has to be easy to do in the first place.

By the way, does not the 'easy' stuff resemble the Core Intent statement? Not surprisingly, the Core Intent is supposed to be easy to use in everyday work.

Besides easy, we can go a long way by moving sustainability towards all the other activities we do in business – the daily operations of a construction company.

Can sustainability help with controlling costs?

Yes, of course, sustainability is about managing financial resources effectively. (Forgetting for the moment that sustainability can also be a cost item!)

Can sustainability help with sales?

Yes, if we know what the clients want and so can modify/adapt/develop solutions accordingly.

Can sustainability be part of marketing?

Yes, we can deploy our sustainability focus to help promote our company. We can also use the talent of the marketing colleagues to deliver internal marketing of the sustainability message to the rest of the organisation.

Can sustainability be part of operations?

Yes, sustainability is part of operations (the rest of the book is about this).

Can sustainability be part of human resources?

Yes, the Lovelace List in Chapters 13 and 14 shows how HR policy and sustainability are linked.

To win corporate commitment, we need to make sustainability a part of daily operations, not an afterthought or an add-on. This means sustainability specialists need to understand sustainability from a business operations viewpoint, not just from a legislative or environmental management viewpoint. Internal secondments from operating teams to the sustainability team and having a business professional running the sustainability function with technical specialists are all ways to ensure that sustainability is not in its own corporate silo (or ghetto) but actually helping every team in the company.

By putting sustainability in the right context with the company's operations, it can be made 'easier'. The personal experiences from the sustainability specialists can help make the changes less difficult and help develop the necessary training.

With the top management seriously interested, I propose the adoption and adaptation of Lean operations as an initial step towards sustainability since they share a great deal of similarity in terms of reducing costs, engaging people and reducing waste. It also fulfils the criteria of making things easier by bringing in a world class approach with the extra bonus that it is also known for reducing costs (see Figure 11.3).

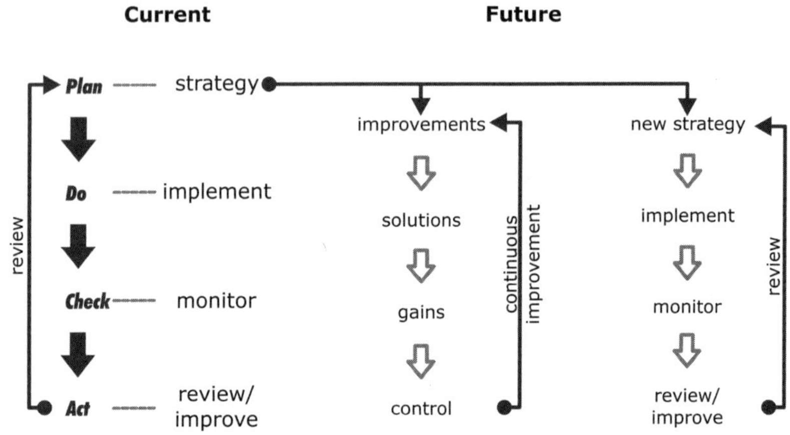

Figure 11.3 Going forward: the PDCA cycle

Managing Strategic Needs

All businesses need to manage strategic needs, but is identifying strategic sustainability needs different from identifying other strategic needs? Although all the regular concerns about understanding sustainability, discussed in the first part of this book, will also apply to how your clients see sustainability strategically, there is likely to be quite significant differences between your views as well as how they see sustainability and how that can be achieved. A developer will have different demands on fulfilling sustainability to a designer or a builder. These demands may again differ with the teams managing the building thereafter. Government agencies and industry bodies, research organisations will also envisage fulfilling sustainability in different ways. The following is a simple example showing the differing factors that affect each group's strategic views on sustainability.

Table 11.1 Factors affecting the differing strategic views on sustainability

Developer	Social needs balanced with profitability
Designer	Balancing requirements: functionality; aesthetics; originality; etc.
Builder	Cost, schedule, specification
Building manager	Ease of maintenance, cost effectiveness, security and social demands
Government*	Fit with overall strategy, balancing conflicting demands
Industry groups*	Promotion of sector, image, social demands
Research organisations*	Originality, gains in reputation
NGOs*	Do good, gains in reputation

* These groups are not necessarily directly involved with the actual construction activities.

Having an accurate picture of the client's needs is equally valid whether a construction company is looking at improvements or transformation. Basically, you survey the relevant stakeholders, map their requirements and then analyse the results. Like a lot of things, it is easy to call but much harder to do. Missing the target in either areas can result in serious mistakes. This is where partnership relations are important in providing a deeper understanding of client needs.

Various tools are suggested in the following chapters to explore client needs: Sustainability Activity Analysis (see Chapter 12); from the House of

Sustainability (see Chapter 17); and the Critical to Sustainability analysis model (see the Introduction).

Review, Update or Upgrade Capabilities

Mapping client needs results in a set of priorities and it is then up to the company to deliver solutions. Like many things, there is an easier way of doing it and there is also a harder way. The easiest way to build up expertise and capability is to upgrade by buying or subcontracting these elements when we need them. This upgrading approach is not unfamiliar to our industry as we subcontract just about every activity in the construction business process. Whether this is the best way forward for your company is another matter. Unlike putting up a building, this is about strategic competitiveness. Do you really want to subcontract the unique elements that make up your company's response to client needs in sustainability? Not subcontracting brings the other, less easy solution – develop the talent in-house by updating its talent.

In Chapter 5, I suggested that a person's capability is made up of initiative, problem-solving skills, knowledge and experience and a professional's green capabilities comprises compliance with the law, minimising environmental management and reducing waste.

A company's capability is much the same except you can adds tools and techniques, equipment and money (see Figure 11.4). Unlike a person's capability which is unique, some parts of a company's capability are common: money; and equipment. Any company with sufficient funds can upgrade their kit to the same level as yours. Similarly, companies with more money tend to have more options than yours. The uniquely competitive factors come down to people, tools and techniques: how a company deploys the available resources.

After dealing with strategic needs and making a decision to either upgrade or update your capabilities, it is time to go forward and make it happen. There are two areas of implementation: improvements; and transformation. The former is about changing the way things are done in the short and medium term, the latter is about transforming the company from 'business as before' to 'business as it should be'.

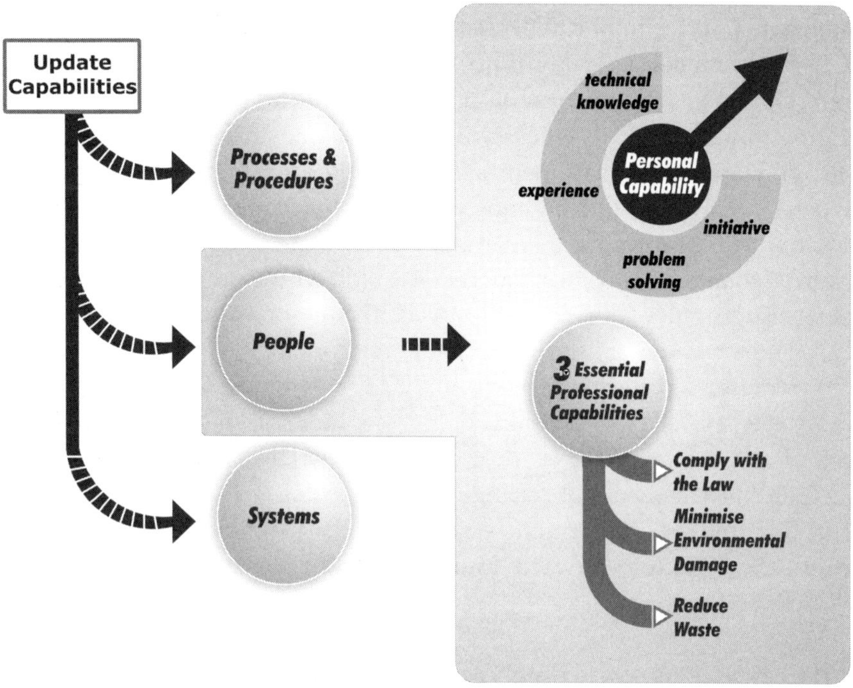

Figure 11.4 An updating approach to developing capacity

Improvements and Transformation

Whatever your plans, the path forward is going to resemble the following steps.

1. Agree a common corporate agenda for going forward and get buy-in.

2. Assess current processes and procedures.

3. Improve the current business processes.

4. Transform the company.

My recommendations (in Figure 11.5) are to follow the sequence to making improvements before considering transformation. There are two reason for this. The first is that through improvements, you will achieve cost savings which will boost your reputation and also temper any corporate fears about how much transformation will costs. The second reason is that by carrying out improvements across the company, you will be able to build-up a core of people skilled in coming up with new ideas and carrying out different ways of doing things. Both of which will make the transformation easier. In fact, once you start the improvement activities, you are starting to transform the company somewhat, and perhaps by stealth. Of course, you are welcomed to start with transforming the company immediately, but the proposed way forward that allows maximum success is to follow the numerical order on the previous page.

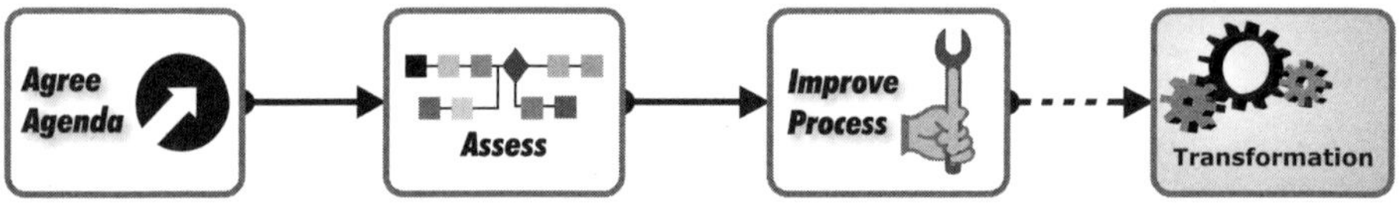

Figure 11.5 The way forward: four step plan

The rest of this book is about these four activities as follows:

- Chapters 12 and 13: assessing the company and its activities;
- Chapters 14 and 16: improving the company; and
- Chapter 17: transforming the company.

Agree a Common Corporate Agenda

In principle this is a simple task. 'Do sustainability' is a fairly good theme for a mission statement and then we comply with legislation; do-good with our neighbours and community as well as training our people. And, oh yes, have some secure bike sheds so people can bike into work. Job done! Or is it?

Not so fast. What does 'doing sustainability' mean for the person pouring concrete? What about the driver washing the wheels of his vehicle before leaving the site? How about the team of designers looking for an aesthetic solution for a floor covering? What does it mean for the planner trying to balance the need for housing and the demand on services? It is not easy for an executive or specialist to come up with an answer to all these questions. It is even harder to encapsulate it in a snappy mission statement. No wonder people hire expensive marketing consultants to do it for them. But that is not really the answer either. Since marketing consultants know little about the operational needs of the operative, the designer or the planner, so they are essentially just trying to repackage your message into slicker words.

Agreeing a corporate agenda is about partnerships with employees, suppliers and clients,[7] it is about agreeing a need to take the sustainability route for real and not for greenwash reasons. It is about senior executives giving real support, middle managers facilitating and participating, and the workforce being engaged. It is about a commitment for the future because the future is about being sustainable.

A simple place to start is to ask ourselves: 'what do we want to achieve?'

Answering this question will lead to a whole range of objectives or 'things to do'. We have now developed a list of aspirations. This is tied not only to client needs but also to internal corporate needs too. The Criticality Analysis (see the Introduction) is a useful way to assess the relative effectiveness of the various objective areas.

The next stage is to ask: 'how will we get there?'

This time the answers will lead to a range of options, things we can do to make the dreams of sustainability a reality. Agreeing a corporate agenda cannot be this easy, can it? Well, it is not likely to be *that* easy, but the principle itself *is* that easy (see Figure 11.6). The trick is to consult the right stakeholders and then sort and list their ideas and suggestions. Then analyse this and assessing which items give the biggest return for the investment in money, time and effort.

7 There seems a trend during the 2008–09 recession that partnering and frameworks are being dropped in favour of lower costs from competitive tendering. The problem with getting the lowest costs is that you may get the lowest in other areas too.

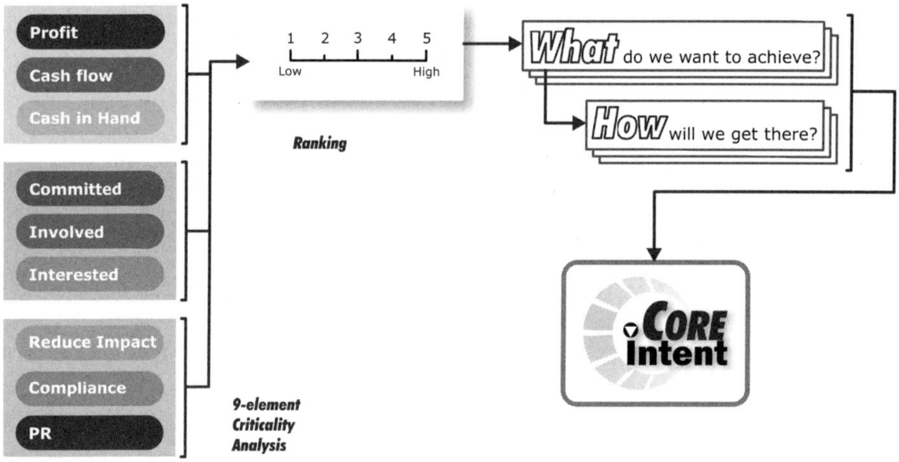

Figure 11.6 Agreeing a common agenda

We can develop a set of corporate sustainability objectives, as discussed in the previous chapter. The critical point about establishing a set of objectives is the weighting factors. These are the factors that shape your particular business directions and separates your company from your competitors. You can use it to take into account the particular needs of the stakeholders. Again we can use the Criticality Analysis.

> **Financial factors**: cost of capital; cash flow; cash in hand; profitability and other financial obligations.

> **Social factors**: needs of business partners and employees; your investors' willingness for risks; demands from both infrastructure partners and community/society.

> **Environmental factors**: impacts of legislation; capacity for reducing waste; and PR opportunities.

Since sustainability is delivered by people, then you want to make sure your people can deliver the sustainability agenda effectively. But does that mean you need to control what people are doing? In Chapter 6, I suggested that empowering people to do their 'thing' on sustainability is better than controlling them but the worry, of course, is the possible loss of the corporate message. And that is where the Core Intent statement comes in.

Deploying the Core Intent

The simplicity of the Core Intent statement approach compares favourably with many existing sustainability statements. The London 4Rs, for example, shows that the four points, although all very inspiring, lacked the concreteness that helps people make decisions. It also looked a bit contrived and therefore did not pass the Core Intent's plausibility test and although it is made up of one-word messages, it is not really that simple. The UK government's Five Shared Principles of Sustainability have different issues: although wordier than the 4Rs, they do not necessarily feel real (in a real-world sense) nor are they simple. Whether they inspire is up to you. Personally, I can hardly remember them except for the 'using sound science responsibly' principle and I probably remembered it wrongly – I do not think it is about warning the mad scientists among us not to try to take over the world, or designing or building resource hungry structures just because we can. I guess maybe these messages were not really intended to help people make decisions but instead, good policy and also grand PR sound bites.

In an increasingly uncertain world, a construction company needs to be making the right decisions that work from the executive suite down to the construction site. A Core Intent statement helps decision making while setting a perimeter for empowered activities. People making effective decisions in new or unfamiliar circumstances can be a source of competitive advantage. It is possible to envisage every potential sustainability activity can be checked by the individual on the spot against the Core Intent for validity.

The Core Intent statement is designed to lead to the following chain of events.

'Will doing this help us to realise (the Core Intent)?'

Asking becomes checking ...

Checking brings validity ...

Validity makes thing happen ...

Making things happen brings results and gives ownership ...

and Sustainability goes forward!

However, without a culture of training, empowerment and encouragement for improvements and innovations, the Core Intent statement will fade on the wall just like all the other mission statements. It then becomes the stupefying straitjacket where the individual resembles a robot, not doing a lot but appearing busy (creating waste?) and forever muttering the Core Intent like a mantra.

The risks of the Core Intent statement ending up like this come from a number of sources:

> *If the company is not ready to receive innovations, then a Core Intent will not do a lot of good. Similarly, if open communications and empowerment are not part of the corporate culture, then the Core Intent statement will just gather dust. In rigid and hierarchical organisations where only managers are allowed to have ideas, a Core Intent statement will appear exotic and probably unwelcomed.*

To make the Core Intent work as designed, there may be a need for a company to go backwards or sideways if necessary to build-up some infrastructure for empowerment and innovations first.

Developing a Core Intent statement can be difficult especially if you do not deal with some potential pitfalls.

> **Political correctness**: a Core Intent statement is supposed to be emotional and fairly short and to the point. However, this can mean that it hits the political correctness buffers since some group's views are not included chapter and verse. By including everything everyone wanted to include can mean a (non-)Core Intent statement that is neither memorable, easy to remember or useful.

> **Lack of understanding about the corporate objective**: a construction company is here to build things, an architect to design buildings. If we lose sight of the objectives, then the Core Intent statement can become dislocated from the realities of the business – very inspiring but having little to do with the company.

> **Not open to everyone**: if the chief executive decides what should be the Core Intent statement then it expresses just one person's intent. Clearly getting other stakeholders' views will cost time and money. But then that is part of the package to motivate people to make sustainability happen.

There is one more potential pitfall which is a bit embarrassing to describe, and it is:

> **Too simple**: there are colleagues who are only happy when presented with impenetrable technical details or paragraphs of words. A simple statement like, 'Reduce waste in everything we do' will probably get confused looks these people – there is not enough detail to help them do the one specific job they have in mind. They are probably the sort of people who wanted to design a specific set of sustainability skills for every single trade rather than having a generic set of sustainability capability. My view is that these people are probably so overworked they cannot think outside their immediate problem (a gentle plug for the problem-solving procedures – full of technical details in the CD).

It is not easy to create a broad based yet focused statement that is fit for every occasion. (This could be the reason why when we see a good Core Intent statement, they are quite memorable.) The difficulty comes from crafting the statement such that it actually does the job and also not have any unintended consequences. I believe the way to avoid a lot of the problems may be to keep the statement simple, such as 'reduce waste in everything we do'.[8]

Analysing the Current Business Situation

The business analysis is a central part in any programme to improve sustainability performance. Although most companies will assess sustainability and waste reduction in their main or core business processes, the actual savings and improvements are also boosted by looking at how to improve the support and back office processes too. More importantly, however, is that the analysis needs to take into account both the demand and supply issues. Again, this is not too removed from the original capability analysis, but this is at a lower level than the corporate capabilities discussed earlier in this chapter.

Demand is what your clients expect from your organisation. This may not be what you can supply in terms of your capabilities but never mind, this is what your clients want from you. The client's needs and expectations establish

8 Even then, unexpected things can happen. In a factory where I organised a 5S waste tagging exercise, a manager was 'red tagged' as waste. This led to the individual having to explain his role and contributions to the overall scheme of things. Come to think of it, it's not a bad idea – we all know somebody like that, don't we?

the demand and it is often more than just delivering a concrete slab or a building. It may also include quality of the finish, relationship with neighbours, compliance with legislation and so on. The list can be quite long for prestige projects and somewhat shorter for a home refurbishment. The demands also shape what you supply. Your organisation delivers the goods as per customer specifications and you need to make sure your teams have the expertise and competence to deliver to specification, on time and to cost. What you supply is your capabilities and you tailor your processes to suit. A simple way to look at this systematically is to use a 'what–how' analysis (see Figure 11.7).

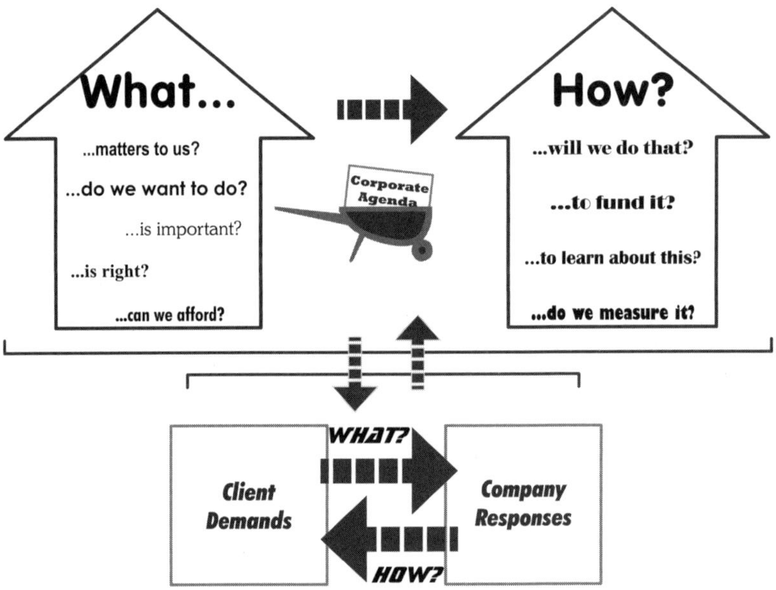

Figure 11.7 What? How? analysis

The client needs or demands are the 'whats' and your responses are the 'hows'. A simple way to manage the demands is by using a balanced scorecard-type model or the Sustainability Activity Analysis Map (see Figure 12.2 in the next chapter) to determine your response. This type of analysis is probably most suitable for a new project when you wanted to tailor a response that matches the demands precisely. For existing companies and existing projects, you can map both sets of realities (demand and supply), then look for bottlenecks in supplying side and then simply address those points. These analyses are described in Chapters 12 and 17.

Only when you are clear about the demands on your services and how you respond to those demands are you in a position to explore the business processes with any degree of certainty.

12

Doing It!

Managing sustainability is something that all of us, whether we are construction managers, designers, QSs, or sustainability specialists, will be doing as part of our professional career. Sustainability is part of everyday management, or will be sooner than you think. Currently we are mostly relying on sustainability specialists and environmental managers to do sustainability for us, but even now, the thinking is shifting. The responsibility for doing sustainability: coming up with ideas on what to do and then implementing, is spreading across the organisation. The CoRE[1] programme in 2008 engaged the construction unions UCATT and Unite and ran workshops in the eastern region where construction operatives discussed the construction waste issues at their workplaces and developed solutions which they then implemented. Similarly, my 2010–11 implementation of the NoWaste programme at the two big Stratford project are about pushing forward with the same agenda (see Figure 12.1). To me, this is the way forward where sustainability is everyone's business and the specialists act as coaches and facilitators.

In the 'talking about sustainability' part of this book, I advocated that sustainability is part of everyone's job and everyone's responsibility. I also suggested earlier that we should deploy process efficiency tools to help us manage sustainability. This part of the book is about managing sustainability as part of the business agenda, much like finance, sales or marketing, except this is across the entire organisation. Sustainability cannot be limited to construction projects or something for the procurement team to impose on suppliers. It starts at both the top and at the bottom. Doing sustainability is a total effort by everyone. And it has to be designed to include everyone, the trick, according to David Farebrother of Land Securities, a developer, is to make it easy for people to do the right thing and make doing it a habit. We therefore need to design

1 The CoRE programme (Construction Resource Efficiency) was managed by BRE and funded through the Learning and Skills Council as well as various Regional Development Agencies in England.

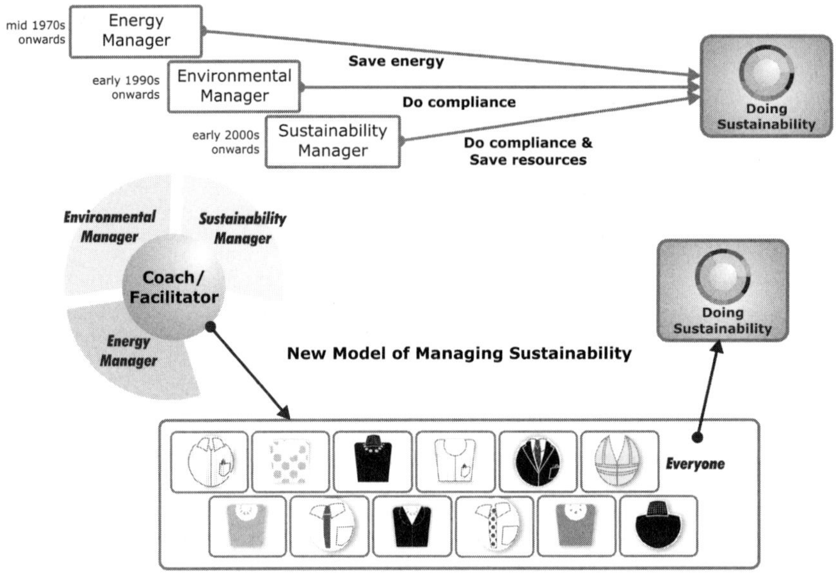

Figure 12.1 A traditional and new models of managing sustainability

our programmes such that everyone (or as many as realistically possible) can take part.

However, before we can rush off and do sustainability we need to have some idea what we are supposed to be doing. Sure, we will have the legislation to provide some guidelines, the Core Intent statement to provide the inspiration, and the 'Eight Wastes' to deal with both in the office and at the construction sites. However, we also need to remain competitive so we need to analyse the sort of activities that we are doing and how these can be managed effectively. Once we dealt with the current activities we can then move on to looking at ways to improve them and deliver improved resource use.

'Doing' Sustainability

It is time to get started: making sustainability happen – reduce our costs but still engaging our colleagues and stakeholders, and yet managing to reduce environmental impacts. There is a lot we can do under this broad topic, but not all activities are the same, nor do they deliver the comparable levels of impact. Some activities require a lot of effort but deliver a low impact whereas others need only a little work but provide massive gains. Since most sustainability professionals and managers in construction are often under-resourced, it is therefore important to consider the values delivered by the various sustainability activity options available.

Analysing Sustainability Actions

Conventional thinking sometimes suggests doing sustainability is basically a combination of legal compliance, adherence to voluntary codes of practice, environmental management and Corporate Social Responsibility (CSR). This book instead suggests a bit more: managing sustainability in construction is a main stream business activity with all the usual product/service management parameters. Therefore, there are two points to consider: what should we do; and how should we do it.

In my opinion, what we can do falls into three areas as shown on Figure 12.2. The Activity Analysis is based on the Kano Model of Customer Satisfaction.[2] The Kano Model splits the things we do into three categories of activities.

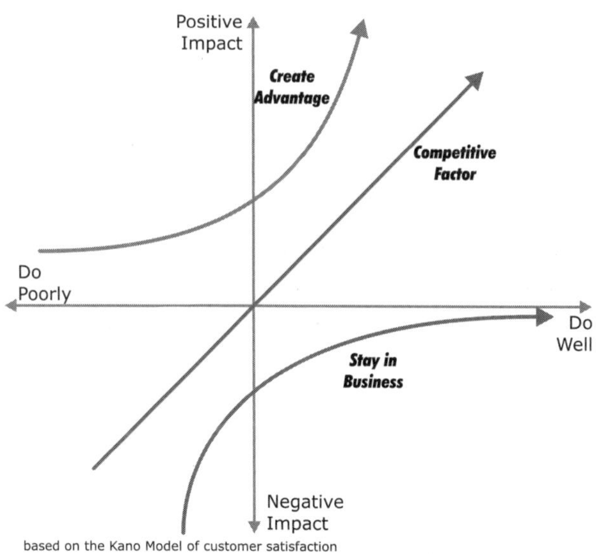

Figure 12.2 The activity analysis map

2 This was developed by Noriaki Kano who suggested that customer requirements often falls into three groups: satisfiers; dissatisfiers; and delighters. Satisfiers are product or service attributes that customers usually ask for and satisfaction increases with increased deployment. Dissatisfiers are something customers will miss if they are not here, although these are not always asked for. Delighters are attributes that the customers have not imagined and are thrilled if they become available.

1. ACTIVITIES THAT ALLOWS US TO STAY IN BUSINESS

These are mainly simple 'do/not do' situation. People may not notice if you do them, but if they are not done or not done to a certain level of competence, then everyone will notice and you will collect a packet of problems.

2. ACTIVITIES THAT ARE COMPETITIVE FACTORS

These activities have a performance element, satisfaction will increase if these are done well. Conversely, if these are not done at all, there will be a certain amount of dissatisfaction. These activities allows us to demonstrate our competitiveness and tend to be classified as general sustainability practices, which some companies do better than others.

3. ACTIVITIES THAT BUILDS A COMPETITIVE ADVANTAGE

These activities are unexpected by most people, consequently they provide a sense of excitement or surprise when they happen. With the surprise factor, it is possible to achieve substantial satisfaction and positive impact even if you only do a little or not being implemented so well (there is no real comparison since this is a novel and surprising activity).

Clearly, with time, any advantage from these activities will erode as the novel becomes industry practice. Early pioneers in sustainability have noticed that their novel ideas in the early 1990s have become industry practices by the 2000s and probably become legal requirements by the 2010s. A good example is the energy saving light bulb, from an expensive novelty through broadening social acceptance and soon to be a main form of illumination.

This is demonstrated by a downward trend in the Sustainability Activity Value Analysis map as shown in Figure 12.3. There is pressure on the sustainability professionals to maintain their leadership position by regularly coming up with something that excites the industry, its clients and stakeholders.

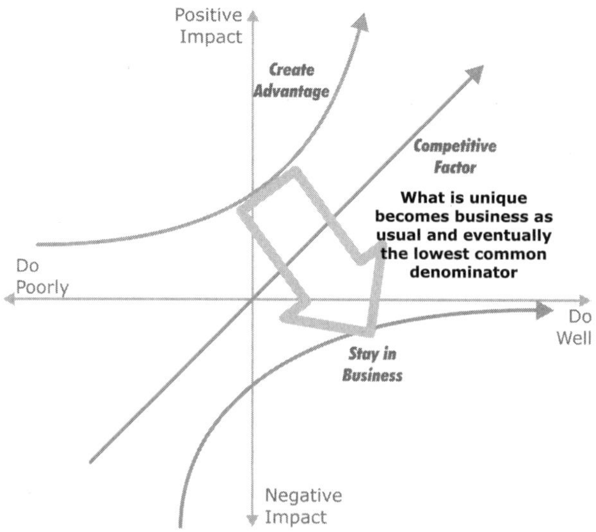

Figure 12.3 The effect of time on the sustainability activity value

Staying in Business Activities

This type of activities include activities that have been around long before sustainability. However, items, such as health and safety, pollution prevention, workforce welfare and so on, can be grouped readily into the social and environmental elements of sustainability. These activities tend to be cost centres rather than profit centres but because doing these keeps your company in business, they are often seen as a necessary business cost.

The Stay in Business activities generally include compliance with legislation. Adhering to legislation, such as the Health and Safety at Work Act or Construction Design and Management Regulations (CDM), will avoid being shut down. However, since it is a matter of comply/not comply, exceeding the compliance requirements does not necessarily bring a higher level of positive impact. Beyond a threshold level, extra efforts are unlikely to gain you more kudos.[3] Nevertheless, drop beyond a certain lower threshold level, then you are likely to incur significant wraith and possibly sanctions from the authorities and society.

3 Health and Safety can be an exception since doing more can contribute to a lower accident risk, but there is still a practical limit after which additional investments brings diminishing returns. This then leads to an ethical dilemma about the monetary value of people's well-being.

A good example is SWMP – the Site Waste Management Plans (see Chapter 10) you do it badly and you may be fined, but no gold star if you do it better than any other company in the UK.

These activities are necessary and very worthy, but they can be hard work when you are setting them up or managing them. This is because a lot of the work is about accurate data recording and checking for compliance purposes. These usually tend to be fairly routine, repetitive and dull. Although to be honest, you really do not want excitement in carrying out the various Stay in Business-type activities: nobody wants accidents or being prosecuted for non-compliance – even though these bring some (negative) excitement.

Managing the Stay in Business Activities

To begin, you need to be able to spot the Stay in Business activities. There are often hints, such as:

- They are on the agenda at site inductions.
- There are posters and charts (from outside bodies like the Health and Safety Executive) about them around the workplace.
- They are on the agenda at many team meetings.
- Individuals are often assigned to manage these activities with special titles (such as safety officer).
- There is often specialist training.
- There are lots of paperwork.
- They are talked about quite openly between competing companies.

Since the nature of the Stay in Business activities often involves regular and accurate data logging, effective management becomes extremely important. It can be readily managed by using a tool, such as the PDCA (Deming) cycle for continuous improvement (see Figure 12.4).

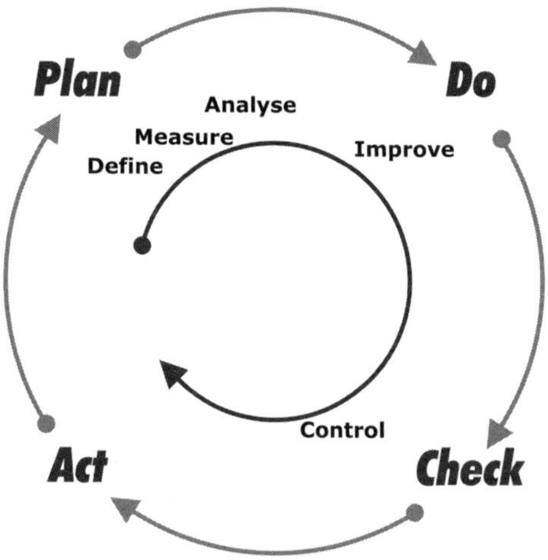

Figure 12.4 The Deming Cycle and Six Sigma

In recent years, this venerable model is supplemented by the Six Sigma equivalent tool: DMAIC (Define – Measure – Analyse – Improve – Control). This is discussed in Chapter 15 and in the CD.

Plan:

- Design a low maintenance but thorough process with straightforward procedure that requires low to moderate levels of training to operate and is robust enough that less experienced staff cannot make mistakes easily.
- Such a process should be designed with clear authorities, responsibilities and reporting procedures.
- The process has adequate testing prior to launch.

Do:

- Make sure that diligence and conscientiousness are personality traits which are part of the selection criteria of the day-to-day management team.

- Encourage continuous improvement from all stakeholders of the particular process.
- There is some form of backup process in case of system failures.
- Provision for sufficient data storage, analysis and reporting.

Check:

- There is an early warning system to highlight potential problems.
- Regular assessment of performance to spot any one-off problems or declining trends.

Act:

- Deploy a rapid decision-making process, such that changes can be implemented easily and quickly.
- Sufficient resources are available to provide training to support any process improvement.

Since compliance is often a major reason for doing these activities rather than gaining a competitive edge, then a realistic strategy for Stay in Business activities is to look for small incremental improvements. Thus regular reviews and inviting contractors to suggest better practice are coupled with regular assessment of the process to look for bottlenecks and resolving them.

Competitive Factor Activities

Unlike the Stay in Business activities, these Competitive Factor activities offer more than the simple 'do/not do' choice as there is also a 'how much to do' choice. There is no limit other than your available resources when dealing with these activities.

This is the part of sustainability that has no end because you can always do more and do better. It can include anything above the legal requirements or standards. For example, the law requires construction sites to provide a rest area that is covered, with tables and chairs as well as something to heat water. The range of on-site welfare facilities we encounter on site stretch from this minimum level to canteens that serve tuna steaks.[4] Essentially, with these activities, the better you do, the more positive the impacts. So if you increase your recycling level, you can appear better than competitors who recycled less than you. Similarly, if you do less than your competitors, you can easily be in a weaker competitive position.

Competitive Factor activities tend to cover the range of conventional environmental management and often include CSR activities.[5] Just about every element in CSR is voluntary but your efforts can be readily compared with those of your competitors.

Be nice to our neighbours?

We can do that better than the project down the road by having quiet hours.

Giving occupational health guidance for your operatives?

We do better by having a therapist visiting our site once a week.

Flexible working?

We do that and also offer everyone remote access for logging on at home.

Value engineering?

Our value engineering process saves more money.

Carbon footprint?

Our analysis starts at where the iron ore was mined, not from the depot in Solihull like our competitors.

And so on and so on. Much of this is about solid progress in doing sustainability, but a certain amount of the effort is also to generate good PR.

4 The Unilever House refurbishment project in London a few years ago had a catering team that served food as good as some of the local restaurants, but at construction canteen prices.

5 Although a recent study seemed to portrait CSR as a potentially ineffective corporate activity – see the *Economist* magazine, 23–29 October 2010.

This can easily become a game of one-upmanship with a very high cost. Although there is nothing wrong with an 'arms race' to do better/more sustainability, there may need to be some caution should this degenerates into simply chasing PR kudos.

Unlike Stay in Business activities where the costs are absolutely necessary. Competitive Factor activities can sometimes offer an uneven return on investment, especially for PR. A few years ago, planting forests is a popular way of offsetting carbon. But once more companies started doing it, the PR impact is lessened.[6]

A subset of the Competitive Factor activities comprises the voluntary compliance activities. These can include the joining of any scheme that requires adherence to a set of external rules: such as ISO 14000; Investors in People; or Considerate Constructors. These activities are not necessary for a company to be successful but they do contribute to the marketing of the company's ethos, both internally to its staff and externally to contractors, clients, investors and society.

Health and Safety is both a Stay in Business activity and also a Competitive Factor activity. Some companies take the former approach and do as little as they need to under the legislation while others make a significant effort to educate, train and support their teams to reduce hazards and accidents.

Managing the Competitive Factor Activities

Since a good implementation in these activities leads to a competitive advantage, it is important to focus on the ones that deliver the most impact for your investment. A House of Sustainability analysis (see Chapter 17 and the CD) is a good starting point to determine the up-to-date client and/or other groups' preferences. From the analysis, you will be able to choose the more effective activities and then it is about improving the way you deliver them. Doing sustainability then becomes a case of continuous improvement and also benchmarking your competitors' progress.

6 The actual do-good impacts do not lessen just because more people are doing it, but unless companies start comparing acreage or tonnes of carbon absorbed by these corporate forests, the feel-good impact of such off-setting can be lessened.

Since most of the high impact activities are well known: carbon footprint; life cycle analysis; better welfare facilities; green purchasing; and so on, it is difficult to satisfy your stakeholder through novelty or uniqueness of your actions. The competitiveness from your activities will have to come from the following:

Plan:

- You need to design or develop a 'better' process than your competitors to deliver the necessary impact. The planning will also need to take into account the cost of training your teams to operate that process.

Do:

- Do it better – for example, your aim can be to reduce wastes in existing process, measure the gains in effectiveness and make sure the information is used to promote your company to the targeted clients and stakeholders.

Check:

- You need an effective monitoring system to spot problems, but it needs to be easy and not too costly to operate. This needs to be matched with sufficient resources to analyse the data gathered within a time span that allows action to be taken.

- To stay ahead, you need to consider continuous improvements.

Act:

- This is having the necessary authority to change the process to eliminate the problems and also to update the skills of the teams operating the process.

Probably the best way to treat Competitive Factor activities is not to consider them as special cases – accept them as part of the corporate competitive arsenal and then assessed and scrutinised in the same way as any other corporate initiative. Once agreement is reached, everyone pull together to make it happen.

Create Advantage Activities

This group of activities are the most difficult to explain: the uninvented ones remain unknown, everything else has gone from being an advantage through to Competitive Factor or Stay in Business activity. But these activities are a blend of your client's needs, your imagination and balanced by your capabilities. The good news is that you do not need to implement the surprise or excitement activity particularly well – the advantage of being first is that there are no competitive benchmarks.[7]

I recall myself being excited when Val Lowman told me of Bovis Lend Lease's work with the long-term unemployed during the building of the Bluewater shopping complex. I was similarly surprised to learn how Laing O'Rourke's Jim Gaffney and Stephen Thomson were engaging prisoners.[8] When I heard of Andrew Pears' Cafe Van programme of supporting small construction projects with sustainability training, there was a similar sense of excitement. I remember sitting at the Building Magazine's Sustainability Awards in 2007 and thinking how some of the finalists were presenting exciting and surprising entries. I felt then that these companies were ahead in both their thinking and doing.[9]

Other than from awards ceremony, where do we find out about exciting and surprising activities? Well, since these activities are also creating a competitive advantage, it is unlikely that lots of details will be forthcoming. There will be reporting in the trade press and you will need to analyse them and then think about what these projects actually managed, achieved as well as what your company will need to deliver in addition to the usual competitive performance – the extra twist that adds the excitement.

I will illustrate with two examples of excitement: the award-winning NoWaste programme I developed and Kotuku's Cafe Van with Andrew Pears. The NoWaste programme is described further in Appendix 1 and the CD, with the following being more of an introduction.

7 When I first introduced workforce engagement to reduce waste in 1996, it was a new idea for many UK businesses. There was so much excitement with this new approach that two of the client team members received MBEs in the 1999 New Year Honours!

8 These were activities that surprised me, my colleagues may not be the first practitioners in these areas and there are many others who do similar good work that perhaps surprised you.

9 A small confession here – Bovis Lend Lease won an award that night with the NoWaste project I developed as a result of that first meeting with Val Lowman.

The NoWaste programme was inspired by Val Lowman's description of her activities at Bluewater, a shopping complex in Kent. It developed initially as a way to provide unskilled workers with a recognisable skill as well as some additional pride. The latter was important because these operatives were responsible for keeping the construction site tidy (i.e. doing the sweeping up) as well as moving materials and removing wastes.

I developed a programme that is based on the Lean workplace organisation process of 5S (see Chapter 15 about 5S) as well as linking this type of work to an National Vocation Qualification (NVQ) award. This provided something that is both new and unusual which generated surprise and excitement.

Some serious adjustments of the world class practices to fit both the actual work and the NVQ was required. (Remember since it is a novel idea, you do not have to do *that* well to get people interested.) These two elements made the programme surprising because previously sweeping up at a construction site is never considered as anything particularly serious. Further excitement came from my using cartoon characters to describe the Eight Common Wastes on a construction site. (And thus brought in the 'Eight Wastes' concept as well – see Chapter 8.)

The final element of the excitement and surprise was to introduce the entire package as a Tool Box Talk (TBT). Unlike the convention ones where the operatives listen to their supervisor telling them what to do and what not to do, these TBTs are dynamic problem-solving sessions where the team and their supervisor discuss a particular waste problem they faced:

> *Supervisor*: (pointing to the poster of Waste from Waiting). We had this last week, why do you think it happened and what is the result of the waiting?
>
> *Operative 1*: Because the lads on level 6 didn't tell us they were going to do XYZ …
>
> *Supervisor*: So it was a case of people forgetting? Or is it about the way we work with other teams?
>
> *Operative 2*: Probably because they have to fill in a whole form just to get a message to us ...

Operative 3: Can't we have a simple card or something instead of the form?

Supervisor: Okay, what should be on the card then?

The TBT also included trainer notes for the supervisors that encourage the team to use world-class tools to solve problems at work. (These TBTs evolved significantly and several are available on the CD.)

So at first glance the formula seems to be: something that delivers the goods but is also unusual, yet simple enough to learn and deploy. Now let us see how the Cafe Van delivers the surprise and excitement.

Andrew Pears of Kotuku runs the Cafe Van, a gas-powered white van that is a travelling training centre. Andrew, who used to run a building company, decided to bring training to construction sites through his Cafe Van and then deliver the training during tea breaks. The aim is to provide a short but informative session to construction workers on waste and sustainability. Unlike other training programmes, Andrew's sessions hit the spot by making sure the context is right for the teams involved. He also feeds his audience rather tasty sandwiches.

There is none of the dreadful monologue from corporate marketers or the techno-babble from sustainability experts. Andrew brings his own bag of 'skip contents' (rubbish to you and me!) and tells it as it is to builders and site managers in a way that leads them to developing their own solutions. All these make up the surprise and excitement factors. But Andrew adds even more – the great thing about the Cafe Van is that it works with smaller builders who are often missed by government training initiatives. Andrew believed that action is only possible when people have the right information so that they can make the necessary decisions. The information needs to be relevant, timely and also something they can use readily. The last point is the major difference between the Cafe Van's efforts and much of the output from conventional programmes. These elements made having the Cafe Van at your project a competitive advantage creating activity.

Besides creating a competitive advantage, the Cafe Van also does good – it often goes to the little known projects (like many of the small projects we see every day when we walk around the neighbourhood) because the teams there probably need the knowledge more than the bigger projects with their corporate training initiatives. Talking to his delegates, I got the impression that he has excited them to think more on the topic and that the majority indicated that their behaviour will be changing.

Andrew's delivery of the Cafe Van programme is professional, clear, straightforward yet a very complex subject is made simple and readily understood. That Andrew also brings along a duffle bag of construction waste (dust and all) also made the training occasion a special one.

I was excited by attending one of Andrew's programmes. My thinking was that this is one advantage that developers, whether house builders or commercial, should not be missing out and neither should local authorities. My excitement extends to a vision of a local partnership of the council, developers and builders having regular Cafe Van visits to the projects in their area. The advantage of a cost-effective training programmes that can change behaviour so effectively at the working level where the waste hits the skips and bins is immense.

Managing Create Advantage Activities

There is a downside to developing innovative, surprising and exciting activities. People will complain because:

- they did not think of it first; or
- thought of it but never did anything with it; or
- they fear change of any sort; or
- they are just jealous.

Remember at the beginning of the book I mentioned the fellow who wrote to the magazine editor saying sustainability is emotional claptrap? This type of person will probably not welcome exciting and surprising activities that lead to a competitive advantage for your company. Others will come up with seemingly considered arguments against any excitement because they fear change. Managing change is hard enough, but dealing with behaviour change is even harder. Getting people to change their behaviour is discussed in Chapter 16.

Introducing novel activities are likely to bring teething problems as well. Again, this can be managed through the PDCA continuous improvement cycle described earlier. To keep the competitive advantage, you need to be focusing on deepening the activities, adding details and elements that keep piling on the satisfaction for your target groups. The House of Sustainability tool (discussed in Chapter 17) is useful in helping you to identify initially the needs and subsequently, the improvements to add more value. The PDCA cycle is applied as follows.

Plan:

- Planning needs to include design and development.
- Map client needs and satisfaction of current products and services as well as competitor activities through a sustainability activity analysis.
- Design new offerings by using tools, such as the House of Sustainability.
- A viable business case needs to be developed.
- Plan out problems by better design and also plan and prepare countermeasures for potential problems.
- Training for teams needs to be in place.
- Select suitable projects for implementation.

Do:

- Implement with client support teams on hand.
- Make sure that marketing and PR activities are targeting the client.

Check:

- Team performance is monitored and assessed against the agreed criteria.
- Check for unexpected problems.

Act:

- Implement solutions and/or updates.
- Control gains.
- Start planning for process improvements.
- Exploring extending activities to other clients.

These are important activities and a senior champion is absolutely necessary, but does the in-house team have the time and energy to develop such ideas if they are managing the entire range of competitive and Stay in Business activities? Outsiders, whether subcontractors or consultants can often offer a broader range of experiences and ideas from other sectors may act as a catalyst for an activity that captures the imagination and provides excitement in the construction sector. Just make sure you get the intellectual property rights agreed before you get started.

Creating Excitement in Sustainability

The two examples highlight how companies can extend their competitiveness through its activities. Excitement may appear to be a unusual topic to go alongside sustainability, but since sustainability is no longer a dry technical affair but one that belongs to everyone, then excitement is another factor that can influence behaviour change (see Figure 12.5).

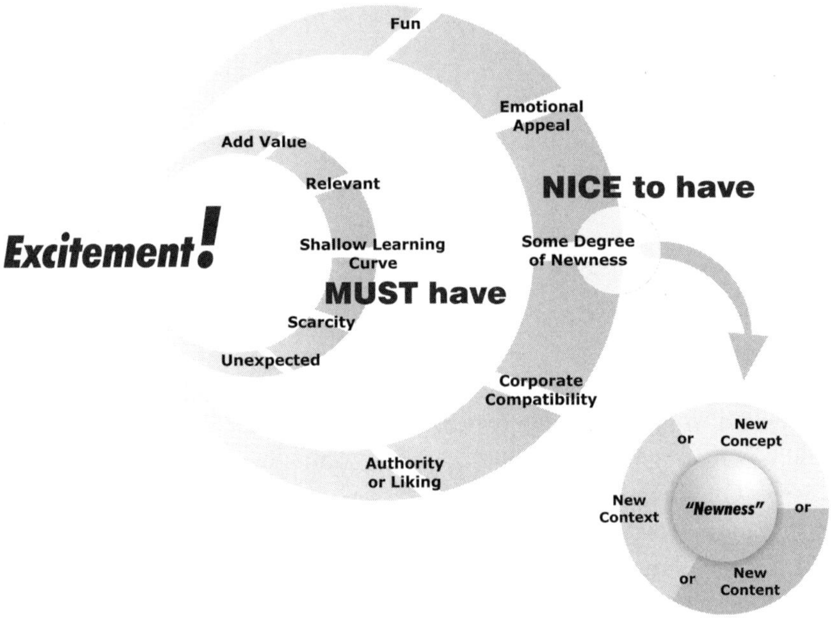

Figure 12.5 Creating excitement: the ten rules

Earlier I discussed the Core Intent message (see Chapter 6) and how it needs to be emotional and appeal to the heart, the excitement factor is also a way to create an emotional bond between the individual and the selected sustainability activity. I suggest a number of rules for excitement, these are:

Must have:

- It must deliver value to the user (skills, capability, increased effectiveness) and fulfil a need.
- It must be relevant to the intended user – context is very important in generating excitement.
- It must also be relatively easy to use/learn/deploy – people will accept a learning curve because of the novelty, but if you make it too hard then you have better make sure the benefits are similarly massive.
- It must not be in common supply – being unique or scarce increases its attraction.

- It must have some element of the unexpected – otherwise it is not going to be surprising, exciting or memorable.

Nice to have:

- It has an element of fun – important for fulfilling people.
- It resonates on an emotional level – people like to believe in something that is important to society, luckily most conventional sustainability activities fall into this category.
- It was developed by someone who is respected or well liked – these attributes adds to the trust for the activity,
- It can readily be linked to existing corporate initiatives – this makes acceptance easier since there is some familiarity.
- It needs to have some degree of newness.

Newness can be an elusive objective since often 'new' can mean expensive. However, if we breakdown what is considered as 'new', then we will find that it may not be as difficult as we first thought.

The 'newness' factor can be split into several components as shown below.

Table 12.1 The Newness factor

Components	Definition	Outcomes
New context	This is using a tested approach or existing data but in new surroundings or with a new group of people. Some adaptation of the data is expected	Accepted as new. Excitement factor: medium to high
New concept	This is a completely new idea or approach. It may contain existing data or being applied in a familiar context	Accepted as new. Excitement factor: high if it is perceived to add value, medium if there is uncertainty about the value
New content	This is about refreshing the details within an existing framework – often this is viewed as a fresh look	Accepted as new if the presentation (concept and context) are new. Accepted as modified if only the content is new. Excitement factor: low to medium unless seen as new – then medium to high

Having all three components being present is therefore not always necessary; it does not have to be completely new: in context; concept; or content. Adapting ideas from other sectors is equally good in generating surprise and excitement, especially if it is based on a proven concept. Of course, if you have all three in your offering and they are all tested, then you are clearly onto a path to gain a significant competitive advantage.

Both the NoWaste and Cafe Van examples above are about novel approaches but there is no need to have an entirely new context, concept and content to create excitement or surprise. Often competitive advantages can come from adapting existing ideas to a new context or putting new content into an existing model, or adapting a process to suit a new context. As long as one of the three is new, there is scope for surprise and excitement, and from there the opportunity to create an advantage.

For example, the House of Sustainability (see Chapter 17) is a case where a concept from a different sector adapted with both new content and new context. Its origin was in product development in ship building and then went to the quality management professionals where it was known as the House of Quality. I took this and adapted it for managing sustainability. The underlying concepts are all well understood and my role was to adapt the specific elements to increase its relevance to sustainability and further tune it for sustainability in construction where it eventually became part of the transformation toolset.

For the Cafe Van, this includes the novelty of the van itself and followed through to the sandwiches and Andrew's bag of construction waste. The fun factor of the NoWaste cartoon figures coupled with problem solving both engage the team and deliver some degree of ownership (see Figure 12.6).

Although both NoWaste and other Lean/Sustainable construction programmes share similar objectives, the more complete and formal implementation in conventional Lean or Sustainable construction programmes meant that it is less easy to learn and has a much lower surprise/unexpected element (see Figure 12.7). To put it more plainly, these programmes often tended to both take themselves too seriously with pages of rules or rigid instructions or are basically exclusive – you needed to be a specialist or a manager to take part.

Figure 12.6 Andrew Pears with his bag of construction waste
Note: copyright Greenfile Developments

Figure 12.7 Overdoing
Source: NoWaste project

The Cafe Van tells you that although sustainability is a serious issue, you can still make it an enjoyable experience. Fun and excitement can be the part of any serious undertaking, it just makes it a more enjoyable and memorable experience.

The unconventional approach by NoWaste, cartoons, problem solving and so on, meant it moved from being a Competitive Factor activity to a Creating Advantage activity.

13

The Lovelace List: A Balanced Social Sustainability Scorecard

Deciding what type of activities we are doing and how they should be managed is a start in engaging everyone in doing sustainability. But that also opens a different area of sustainability, if we want our people to be motivated and taking part in making sustainability happen, then how are we going to prepare them? There is training, updating processes and procedures and so on. But the most important is to motivate our people.

People can be motivated by inspiring words but it is not sustainable in the medium to long term. The transformation of our companies, projects and changing the way we work from 'business as before' to 'business as it needs to be' is not a short-term event. The transformation is a multi-stage affair and there are changes needed not just in techniques and products but most importantly, in helping people to change the way they think and work (see Figure 13.1).

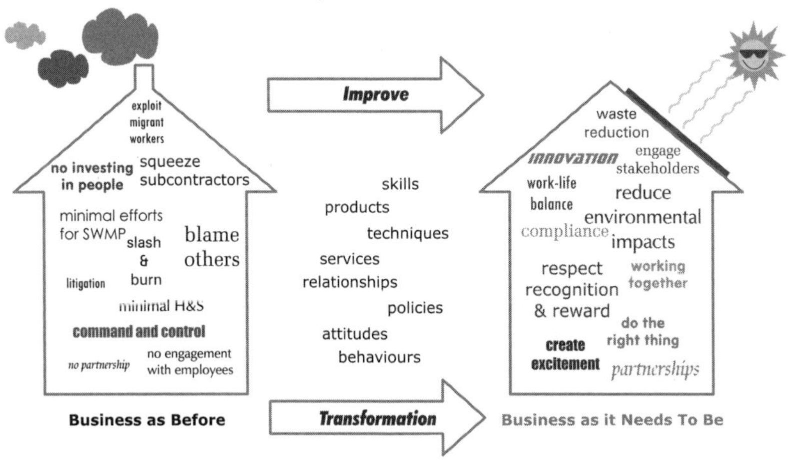

Figure 13.1 The transformation of companies, projects and the way we work

Engaging people brings us back to the social aspects of sustainability.

The Lovelace List

My colleague Rebecca Lovelace helped me to develop the themes in Chapter 4 and these led to a comprehensive checklist on the types of activities that can contribute to a social sustainability programme (see Figure 13.2). This is a balanced scorecard[1] (see Chapter 4) for social sustainability, which I have named the Lovelace List to thank my colleague for her ideas.

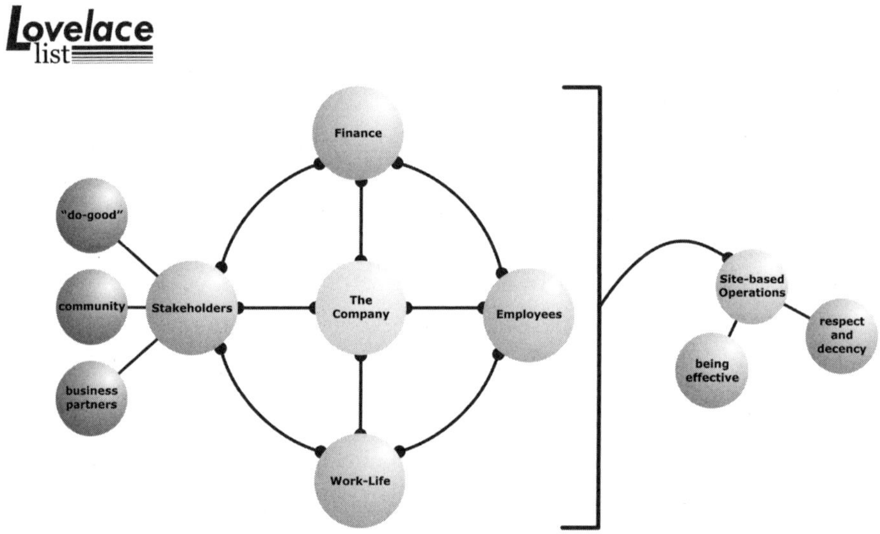

Figure 13.2 The Lovelace List: a Balanced Scorecard for social sustainability

The elements of this scorecard are:

- the company and its finances;
- the company and its employees;
- the company and its attitude to work–life issues;

1 For clarity, the Lovelace List is a 'balanced scorecard', whereas the Kaplin and Norton original is 'The Balanced Scorecard.'

- the company and its stakeholders; and
- the company and its construction site-based operations.

Like other balanced scorecards, the aim is to achieve a balance with all of the elements. Although all the individual elements contribute to a specific aspect of sustainability, over emphasis on any of these will lead to imbalance and a reduction in overall effectiveness. For example, too much emphasis on the skills development of the employees may make the employer seem like a slave driver, forever increasing the capability levels of its people without concern for their personal well-being. Excessive attention on the corporate finances will benefit one area of sustainability, but will also miss other opportunities to enhance corporate capabilities. Similarly, too much 'doing good' may upset the shareholders who may not wanted their investment to be put into such activities.

The Lovelace List is described below and the checklist itself is described in the next chapter and available from the CD.

The Company and Construction Site-Based Activities

The construction site-based elements relate to the site and its operations, and are presented as a quick checklist below:

- We ensure that equal opportunities for everyone is a reality on our project sites.
- We check and update our method statements regularly and for every new project.
- We respect the dignity of our teams on site and our welfare facilities is one of the ways we show this respect.
- We focus on health and safety not because of the law but because we really believe it is important.
- We care about our teams and so we pay serious attention to occupational health issues as well.

- We want everyone on site to be able to work effectively, so we make sure the work is designed to minimise problems, hazards and errors.
- We want our neighbours to be able to enjoy their lives with minimal disturbance from our work, so we make sure the work is designed to minimise problems, hazards and pollution.
- We are ready to listen to ideas from our teams and we support our colleagues who come up with ideas for improvement.
- We keep the site clean, clear and safe to make it is easier for everyone to work.
- We take training seriously and schedule regular Tool Box Talks as well as special ones when there is need.
- We respect our teams for their ideas and innovations in designing out waste, reducing waste, reusing and recycling.

The site-level Lovelace List is made up of many pre-existing activities but joined with a central theme – putting emphasis on the people doing the work and paying attention to how the industry treats people. It is more than likely that many companies are already doing a majority of these things. The checklist has two main areas that contribute to sustainability: decency and respect; and effectiveness. The former is about the dignity of our colleagues and the latter is about addressing our teams' basic needs to deliver a good result.

Site: Respect and Decency Elements

Equal opportunities: this is the law but it can be easily flouted. We are not suggesting that quotas for race, sex and age should apply in all areas whether it is appropriate or not. It is about giving people relevant opportunities to develop where they can fulfil their aspirations. Paying attention to the financial aspects of sustainability requires you to put the right people onto the job, otherwise you end up with political correctness but not a lot of economic value. Social sustainability is about the realistic and thoughtful provision of opportunities.

Welfare facilities: if we treat people poorly, then we can expect them to behave poorly and deliver a poor quality of work. Skimping in these areas is also morally wrong because it disrespects people and that is not a path to sustainability.

Health and safety: from the trade press, it would seem our industry is in a cycle of gaining, slipping and gaining again. Construction is made up of lots of young men believing they are indestructible. But it would seem that either we are often too rushed, not trained sufficiently, not following procedures or some other excuses used to explain safety lapses. Whatever we say, we need to do more, not because of the law but because we are decent people.

Occupational health: this is about the physical, mental and social well-being of workers. This means not only the health and safety aspects of construction and the CDM Regulations, but also general hints, tips and guidance on personal health.

It seems that often healthier living and occupational health are still not taken as seriously as the legislation-driven health and safety at work issues. So it is good to hear that Royal BAM Group was giving their people full health checks and Bovis Lend Lease was offering healthy foods at some of their site canteens (reported variously by *Construction News* in April 2007). I seem to recall hearing that Mace offered foot massages – maybe it is anecdotal but it is a good idea. Many progressive companies in the UK offer guidance on and support for healthier living and there is no reason why every company should not do the same – the information is free from many sources. A healthy worker is a more productive and happier one.

I discussed this with Veronique Minier, a professional massage therapist, and her opinion was that construction is an occupation that often requires lifting loads or twisting in and out of awkward spaces, both of which can put a strain on the muscles and tendons. Minier believed that regular on-site visits by a specialist can help many operatives sort out their pulled muscles and stiff joints at an early stage, thus preventing long-term injuries. She also felt that the Japanese fashion of morning stretches before work starts is a good idea for an industry that depends on the fitness of the teams. Although in a macho industry like ours, Minier did concede that there may be difficulties in accepting some occupational health issues, but as an operative from Ethiopia told me, 'you only have one set of health and safety'. Many of the operatives are self-employed and early treatment, or rather self-preservation (a kind of sustainability, don't you think?) will mean that they get timely treatment so that their ability to work is not threatened by debilitating injuries.

Neighbours: nobody likes noisy or obnoxious neighbours. A construction site is often noisy, dusty and unpleasant to its neighbours. There is not a lot we can do to reduce the physical noise of demolition or operating plant, but we can readily let our neighbours know what is going on, be considerate as to noise, timings and dust reduction, and consult on the best ways to work to minimise the impacts of a construction site. We can also do something about the lack of courtesy on site from the wolf whistles and drooping trousers.

Site: Effectiveness Elements

Method Statements: how often are these really updated to take into account different situations, and different tools and materials? Yesterday's methods may no longer be enough to give us a competitive advantage, it is just like having an ancient map of a flat Earth, you can go only so far. The updating is also an opportunity to get the team to help design out waste by improving procedures and methods.

Work and job design: this is linked to working methods, workplace organisation, health and safety and communications. If we spend some time designing the work, taking into account other activities, we can plan the overall rhythm of the construction project better. This is also the practice of the leading 'Lean constructors' like Shepherd Construction and Thomas Vale.

Continuous improvement: not only does this increases competitiveness, a positive response to innovations also raises morale and encourage more active partnership on a construction project – what is not to like about this?

Training and skills development: by putting relevant training elements into the training plan, CITB[2] registered companies can get rebates towards

2 The levy-raising and training rebate-giving bits is still often called the CITB (Construction Industry Training Board), despite the newer name of ConstructionSkills for the skills and policy bits.

training. Your people become more skilled, you get a discount on your training costs and your competitiveness goes up, again, what is not to like about this, especially as you have already paid your CITB dues?

Linking Decency and Effectiveness

Workplace organisation: can a job be done more effectively in a better working environment? (see Chapter 16 for more details). A neat and tidy working environment and a structured work plan mean there is less need for 'work-rounds' thus minimising the possibility of errors and mistakes.

The Lovelace List – For the Organisation

At the organisational level, the Lovelace List comprises several balanced areas of activities, these are:

- managing employees and their development;
- managing corporate finances;
- working with stakeholders; and
- managing the work–life balance.

Managing Employees

The List outlines the activities employers can undertake to develop their people, not only in gaining new skills, but also in engaging them, empowering them and developing a partnership between the employer and the employees.

- We support our colleagues to adopt lifelong learning and fulfil their aspirations.
- We have programmes to help our colleagues further their development.
- We encourage employees to participate in decisions that affect them.

- Our people feel that they are listened to and receive constructive feedback.
- Our people feel supported by management and are confident in the leadership of our business.
- The values of the business are clear and employees identify with and support them.
- We have a diverse workforce.
- We have a zero tolerance approach to all forms of bullying and harassment.
- Our business is a safe and healthy place in which to work.

These elements represent an understanding that to get the best from people, they need to feel valued: people development; opportunities to grow; and ways of becoming engaged in the company's activities are paramount.

Managing Corporate Finance

This is the caveat of social sustainability, indeed of all the sustainability elements. BATNEEC[3] applies especially in this area. Selecting a pension fund is difficult enough and investing into ethical funds where the returns may not be the highest may conflict with corporate rules on maximum gains in investment. Do not forget, you are dealing with people's future economic well-being. Also, remember BATNEEC – if you cannot afford to do it, then do not! In a recession, ethical investments may appear to have less relevance since interest rates or rates of return are going to be low, but it is often the highly speculative investments that often crashed and burned first (although in the 2009 recession, just about everything trashed).

- We invest in ethical pension funds.

3 Best Available Techniques Not Exceeding Excessive Costs, or do not go for things you cannot afford, but get the best with what you can afford. A get-out clause for some and a dose of realism for others.

- We have a fair and transparent profit-sharing or Gain-Sharing scheme for all employees.

- We pay our employees a fair wage for the work they do.

- Our pension scheme includes a contribution from the employer as well as the employee.

Gain-Sharing is a different approach towards rewards based on agreed productivity gains. This type of rewards system is probably most effective for the entire staff rather than individual employees (see Chapter 16).

Working With Stakeholders

From the Critical to Sustainability Analysis (see Introduction), the stakeholders a company encounter can be divided into three simple groups: the committed (the workforce and investors/owners); the involved (the business partners: suppliers, clients and others with a contractual relationship); and the interested (society as a whole, interest groups, pressure groups and everybody else). These people also form the 'voices' that any construction project needs to take into account nowadays (see Figure 13.3).

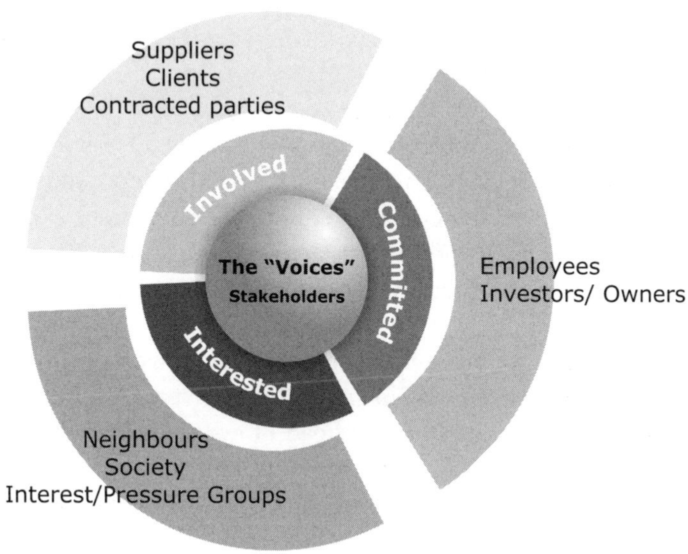

Figure 13.3 The voices: stakeholder groups

STAKEHOLDERS: BUSINESS PARTNERS

There needs to be a greater emphasis on the supply chain as parts of the construction industry often treat their suppliers, especially the smaller ones, poorly. As for clients, since most companies bow and scrape to their customers, there is not a lot more to do, other than clenching your teeth perhaps! Nevertheless, ideas and innovations in sustainability should be shared with both suppliers and clients.

- We respect our supply chain partners, both in words and actions.
- We deal with our supply chain ethically and in a transparent, open way.
- We work together with our supply chain partners in dealing with sustainability issues.
- We encourage and help our business partners to achieve improvements in sustainability.
- Where possible we source goods and services locally.
- We share sustainability ideas with our supply chain partners regularly.

STAKEHOLDERS: DOING GOOD

Doing good is an area where corporate goodwill is not all about satisfying the egos of senior executives. Companies belong to shareholders,[4] and achieve their values from the efforts of the employees. Before giving money (or employee effort) to worthy causes, it may be prudent to secure the agreement of the shareholders, after all, it is their investment (and their company).

Similarly, consulting the employees on which worthy programme to support is also a prudent move. This provides an extra motivational element for the people in your organisation. Instead of just working for the company (and the 'fat cats' in the company), the employees are also helping other people through their efforts. This is especially so if the support for charities is included

4 Although this is the case for every company; owners, investors and shareholders in charities and NGOs are seen differently by those who invest in commercial or profit-making companies.

in a Gain-Share agreement where productivity gains are visibly translated into the support given. Once again: only if your company can afford it. A half-hearted effort can destroy morale very quickly.

- The employees in the company choose the charities we support.
- We support a mix of national and local charities.
- Our employees are given paid time-off to help the community.
- We have a well-established, charity fund-raising programme.
- We support our employees in charity fund-raising by adding to or matching their contributions with a corporate grant.
- We give a percentage of our profits to charity.

STAKEHOLDERS: COMMUNITY SUPPORT AT BOTH CORPORATE AND SITE LEVELS

Community support is a popular item on many corporate agenda, and some may see it as a sop to political correctness and some will consider it an opportunity for self-serving activities. But whatever it is, being a welcomed member of the community enhances the organisation's reputation, and if your community support results in more people being trained in our industry, it then helps to secure the supply of future employees.

The point to remember is that construction companies are not there to deliver social services,[5] but are there to make money through construction activities. However, even in a globalised economy, buildings are physically anchored to one spot and it cannot be outsourced to China. Therefore, since the companies' activities are anchored to the soil of this country, then community support does seem to be some form of investment. But like all investments, if you cannot afford it, do not do it.

Both at the corporate and project site level, there are opportunities to help people progress. The construction industry has several formalised programmes,

5 Although the current UK government is talking about big time social engagement, time will tell whether companies (and individuals) during a sluggish recovery will be willing to release resources towards doing public good.

such as 'BeOnsite' from Lend Lease and 'Building Futures' from Wates. These are wonderful efforts and deserve the support of (and emulation by) the entire industry.

Do you expect to get something back for supporting the community besides goodwill? Well, if it is return on investment in the traditional money sense, this is unlikely. But if we are talking about investing in human potential, doing our part to build a better world, then the answer is an overwhelming yes. Giving people opportunities through community support is one of the better things a company can do.

- We are involved with local schools, communities and other local partners to encourage and support newcomers to our industry.
- We support local communities by offering work experience and educational visits for young people.
- We support local communities by providing opportunities for people from vulnerable and marginalised groups.
- We support local employment and training initiatives by sponsoring individuals towards apprenticeships and training.
- We give our employees paid time off to take part in local mentoring and volunteering schemes.
- We engage with those people living near our site who may be affected by our activities through newsletters, open days, charity and school involvement, etc.
- We aim to leave a strong legacy, as well as a building, behind when we finish construction.

Work–Life Balance

All work and no play is not a sustainable solution.[6] A healthy workforce benefits the employer, as does a happier and more motivated workforce. In

6 Although of course, work takes place at work and play takes place somewhere other than at work!

times of skills shortage, work–life balance benefits do contribute towards staff retention. They also contribute to better relations and understanding between the employer and the workforce. Better understanding may come in useful in economic downturns where cutbacks to benefits may be needed. Of course, a company may deploy a Gain-Share programme with the benefits as part of the share from productivity gains in a tougher economic climate. In a recession, a committed workforce is more likely to innovate whereas a withdrawal of goodwill from both employer and employees over reduced benefits can be counterproductive.

What you choose for your organisation depends on what you can afford with respect to the overall business environment. The following is our recommendations.

- We support flexible working in our company.
- We have a sabbatical scheme for staff to take time off.
- We offer a career break scheme for our staff.
- We do more than the law requires for maternity and paternity policies.
- We provide support for our employees in dealing with stress and well-being issues.
- We provide access to healthy food for our staff, where possible we source this from local providers.
- We support healthy travel to work through the provision of showers and bike sheds.
- We offer an annual medical check-up for our staff.
- We make it possible for staff to work from home, if appropriate.
- Employees are given above average holiday allowances.
- People we employ enjoy working for us.

- We think work should be enjoyable.
- Our company culture recognises the importance of happiness.

Reality Check

The Lovelace List is not set in stone but instead is a living document that will change and evolve as companies progress in their social sustainability activities. It is not exhaustive either but is a list of suggestions that will help you put your organisation on the path to social sustainability. You need to make sure you remember the BATNEEC principle, if you cannot afford it, then do not implement it. Nevertheless, look for items that will give your company the maximum return on investment as well as a set of balanced activities.

Is the Lovelace List a set of KPIs? The Lovelace List is our suggestion of the directions your company can take in social sustainability. It is not a traditional set of key performance indicators with hard targets which does not take into account the individual situation at each company. Sustainability has no real set limits[7] and we can always do better. So, yes, do use these as a set of KPIs, but with a twist – these are continuous improvement KPIs. You get to decide on the activities that will realise your aims.

You are welcome to add extra elements to suit the particular needs of your organisation. If you want to share your suggestions for the Lovelace List, please do so by emailing your suggestions to Gower Publishing, my publisher, and we can add it to the next edition.

Ultra-Compact Summary of Social Sustainability

Social sustainability and the Lovelace List can be abstracted into just three words: respect; recognition; and reward. This is sustainability in an ultra-compacted form. We need to respect the people in our teams for their innovations, ingenuity and the ability to deliver value. We need to recognise these efforts and say more than 'thanks', we need to demonstrate this through a realistic set of rewards. Saying this is easy, but doing this can appear less so. This is discussed in Chapter 16: the 3Rs.

7 Other than those imposed by your resource constraints.

14

Using the Lovelace List

The Lovelace List is a balanced scorecard for social sustainability and it is intended to be used in a balanced fashion where the focus is spread across the entire range of elements: finance; employees; stakeholders; and work–life balance. The List is divided into five main balanced scorecard areas: employees; finance; work–life balance; stakeholders; and site activities. Each List item allows you to assess your company's progress towards social sustainability at five levels or stages.

The Lovelace List can also be deployed as a checklist to assess your company's current progress in a suggested sequence as shown below (see Figure 14.1).

1. self-assessment
2. analysing the corporate policy and workforce needs
3. balancing your activities
4. Class 5 Questions Cluster (C5C) analysis
5. implementation
6. measure progress
7. communications.

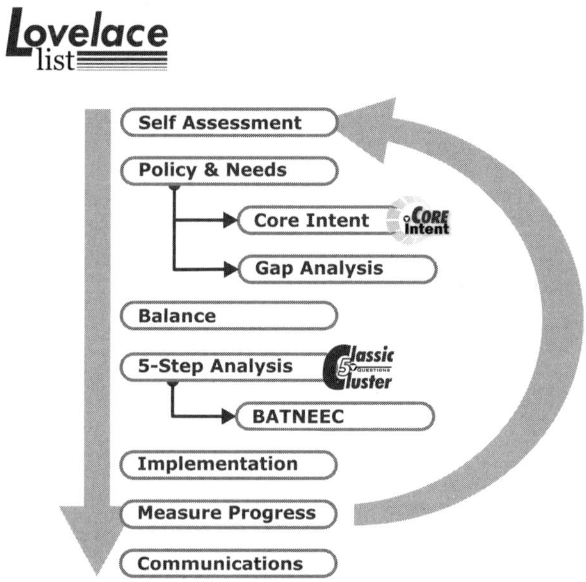

Figure 14.1 The Lovelace List: process sequence

1. Self-Assessment – the Checklist

Assess your company on how closely it matches the statements on the Lovelace List. The scoring system has five different stages of intensity.

There is no right or wrong answer. It is a checklist about your current practices. The different stages may be shaped by enthusiasm, pragmatism or resource constraints. The important thing to remember is that you need something that fits both the needs and the budget of your organisation.

The five stages are described below.

Doing nothing	You have no programmes or activities in this area at all

This may be a money saver but you need to remember that social responsibility is an area of competitiveness nowadays. Your clients may have already asked you what you are doing in this area, your competitors are probably doing some of the items on the Lovelace List already. The impacts of doing nothing go beyond this: possible higher accident rates; greater staff turnover; and a poor reputation. Do not get left behind, this is not a race you can catch-up easily after a slow start.

Doing very little	Some planning may have happened, maybe a team has formed but progress is still very slow

Some companies have a plan of sorts and a set of principles or a vision statement, and that is about it. This may be because social sustainability is seen as a bit 'touchy feely' and so attracts a low priority with minimal budget and staff resources. (Do not believe people who tell you sustainability is free – nothing is free in sustainability, even 'feel good' activities need somebody spending time to set it up.) However, having a social sustainability statement that is not backed up with real action is not really sustainable. You lose out to competitors who have done things and won awards or written case studies. It also makes you vulnerable to major clients wanting to audit your sustainability progress.

Doing some	Your company is reporting some results in this area, perhaps irregularly and maybe depending on the enthusiasm of certain individuals

Many companies are doing 'something' in the social sustainability area. This level of activity is probably a case of doing the best you can under the circumstances. While this is fine in itself and not a bad place to be, you need to remember that your competitors may be doing more. It does not matter whether they are bringing out the smoke and mirrors. If you are not doing the things clients think your competitors are doing then you run a risk of losing reputation and possibly business.

Against this you need to balance what you can do with the budget, time and efforts. So it is a case of doing what is possible and doing it well, rather than do everything but doing it poorly.

Doing all regularly	Your company has an integrated programme and you are meeting your targets and setting new challenges

Not a lot more to be said here, you are well on the way to making sure that your corporate reputation is boosted not only by you being a good contractor, but also being a good employer and business partner. Your organisation has clearly grasped that although social sustainability has a cost, it also results in greater motivation with the workforce, the respect of society and satisfaction with clients and business partners.

Of course, you also realise that any activity can be improved upon and you focus on what you do and how to do it better. It should not be that difficult since your employees are also your partners in the continuous improvement activities (see Continuous Improvement, Chapter 15).

Others learn from us	You do it so well that other companies are asking you how it can be done. You are a leading authority

You share and exchange ideas and experiences with business partners. Mostly you will find that companies in your supply chain and perhaps your clients come to you to learn how it is done.

You also encourage your employees to speak at conferences and seminars both to promote your company (and show off) as well as to learn from others. This is because you realise that no matter how well you are doing, someone else may have an idea you have not thought of. What often happens is that companies outside your sector come and learn from you as well.

Not only do you take your social responsibilities seriously, you also realised that after all the hard work, everyone deserves a pat (or two) on their backs. So you engage your marketing and public relations teams to make sure your efforts are flagged up to potential clients and other partners. This can end up with local radio and TV coming around to interview you and your colleagues and maybe even going to industry events to collect awards for your efforts in sustainability. The awards are a public recognition of your achievements.

After the self-assessment, you will be able to identify areas that you consider as the next steps.

A SLIGHT DIVERSION: IS SHARING BEST PRACTICE A GOOD IDEA?

During my research for this book, a manager asked me why on earth should they share their best practice and lose their competitive advantage? This is a good question and the fear of losing competitiveness is a real one. However, there are two sides to this issue.

If your company does not improve on current practices, then sharing them can lead to a rapid erosion of competitiveness since your competitors can overtake you. Although most companies want to improve, the situation in many companies is that only the managers are allowed to be involved. There is therefore tremendous pressure on these managers to innovate and to do it year on year. There is a limit of what can be achieved with a few willing individuals and the managers tend to run out of ideas after a few years. In these instances, continuous improvement falters and sharing best practice can be a major risk.

Nevertheless, world class companies know better: best practice is a mixture of techniques and company culture. Competitors may be able to copy the individual techniques, but without the people behind them, the copying will be incomplete. By the time your competitors get the technique to work, your company will itself have moved ahead by 6 to 12 months, thus maintaining the competitive lead. World class companies also know that innovations come from everyone and so engage both managers and workforce to come up with better ways of doing things. The other advantage of this approach is that even if a critical manager joins your rival, the majority of your continuous improvement team remains intact.

Jon de Souza of Constructing Excellence, an industry body, suggested pragmatically that sharing sustainability best practice is probably going to be the way many companies get their start in doing sustainability. He felt that there needs to be a more systematic transference of best practice between the leading edge companies and the late adopters. So the message for world class companies is quite clear – get your current best practice case studies to Jon and his colleagues at Constructing Excellence. Then bask in the reflected glory of being a sustainability leader while you develop the next stage of best practice!

2. Analysing the Corporate Policy and Workforce Needs

Now you have an indication of how your company is doing in terms of social sustainability. Is this where your stakeholders want your company to go? You can use the first component of the transformation process (see the Needs Matrix in Chapter 17) to check the demands and needs on your company.

The Core Intent (see Chapter 6) should be integrated into this assessment because it is an outcome of your company's sustainability programme as well. The Core Intent is your company's empowerment message to all its employees and subcontractors. It both encourages individuals to take the initiative yet also provides a perimeter for these endeavours.

The needs analysis outlines what areas require corporate attention and this can be matched against the low scoring areas in the self-assessment checklist, described earlier. Together these offer an indication of where the gaps in current performance are when viewed against both the corporate objectives and the Core Intent.

You can review these readily from the Lovelace List self-assessment checklist and tick-the-boxes on the 'Action Needed' column (see CD). These ticked boxes will provide you with a ready reckoner of what are the priority areas.

It is all very well analysing the corporate policy, but so what? What about the needs of the workforce? My own research, coupled with discussions with Lynn Williams of Lunces Training, a productivity consultancy, showed that workforce needs can be grouped in several areas (see Figure 14.2):

- partnership
- relevant empowerment
- communications
- respect, recognition and rewards, and
- lifelong developments.

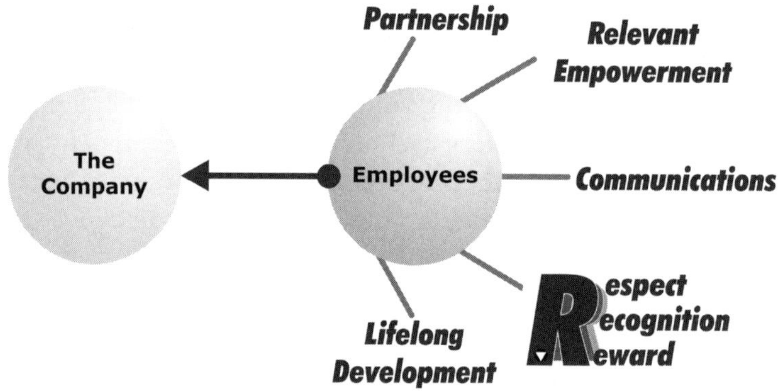

Figure 14.2 Workforce needs

Partnership, in the workforce sense comprises:

Table 14.1 Partnership

Activities	Compatibility with the Lovelace List
• Sharing common goals between management and workforce • Cooperation and working together to succeed • Sharing success • Trust • Equal opportunities • Increased participation • Broader involvement • Common sense/natural justice • No 'us and them' • Job security	The elements of Lovelace List represents a network of partnership, and the workforce needs are covered in both the site operations, stakeholders and the employee areas. The teamwork nature of construction means that partnership is a natural way of working

Empowerment is a highly charged word, but the workforce is not necessarily after *power*. They know very well that there are things that they do not need to know about or control, but they also know that there are things which they are best at doing. It is about relevant empowerment, with a perimeter defined by the Core Intent statement. The issues are readily tied to the Lovelace List.

Table 14.2 Empowerment

Activities	Compatibility with the Lovelace List
• Shop floor influence in local decision making • Relevant recruitment (hiring the right people) • Shop floor has accountability • Involvement in how the work is organised	Construction is regulated in terms of employment, health and safety as well as environmental management. Nevertheless, the Lovelace List area of being effective at the working level means empowerment that is relevant should be explored and adopted where appropriate

Communications has always been a major issue in any organisation and although it is not always intentional, a great deal of corporate communications is sometimes obscure and unhelpful. While this may be fine for general spin and hype, it can distract or confuse the workforce and misdirect them. In terms of sustainability, this can be a major problem. The common communications needs, listed below, offer no surprises to anyone.

Table 14.3 Communication

Activities	Compatibility with the Lovelace List
• Sharing information and communication about the organisation, the work and the directions of the business • Openness in communications • Meaningful and understandable communications	Respect and decency, employees and business partners are areas where the Lovelace List and the needs of the workforce converge. As construction can be a dangerous activity, effective communications is something that cannot be set at a low priority

Respect, recognition and rewards forms a very short summary of what any company's policy on social sustainability should be. If you do these three things well, then you are on the right tracks for social sustainability. The workforce needs in this area is not just about £££ of rewards but also about a degree of decency (more in Chapter 16).

Table 14.4 Rewards and recognition

Activities	Compatibility with the Lovelace List
• Mutual respect and recognition of contributions • Being valued for their opinions • 'No blame' culture • Recognise the achievements of individuals • Open and fair reward system	Our success in construction depends on the efforts of our colleagues around us. The Lovelace List covers respect and decency not just at the construction site but also with employees, the work–life balance as well as in dealing with stakeholders

Lifelong development in construction is a not always easy to achieve since the prevalent business model is subcontracting, this makes it difficult for major contractors to engage in the lifelong development needs of subcontractor staff. However, in partnerships with both the client (yes, they have a responsibility too), subcontractor and other bodies, such as UCATT the union and Construction Skills, this can be achieved to quite an extend.

Table 14.5 Development

Activities	Compatibility with the Lovelace List
• Personal growth	The Lovelace List covers the lifelong development needs of the individuals. Both in professional areas (employees) and in helping out communities (stakeholder/community)

3. Balancing the Activities

Once the corporate needs in this area have been agreed, you should consider whether these are in balance with each other. The arguments for balance is not too different from those I propose for the sustainability elements discussed earlier, but social sustainability does differ in that the impacts happen to people. If the balance is managed poorly, there can be significant morale loss in your teams (see Figure 14.3).

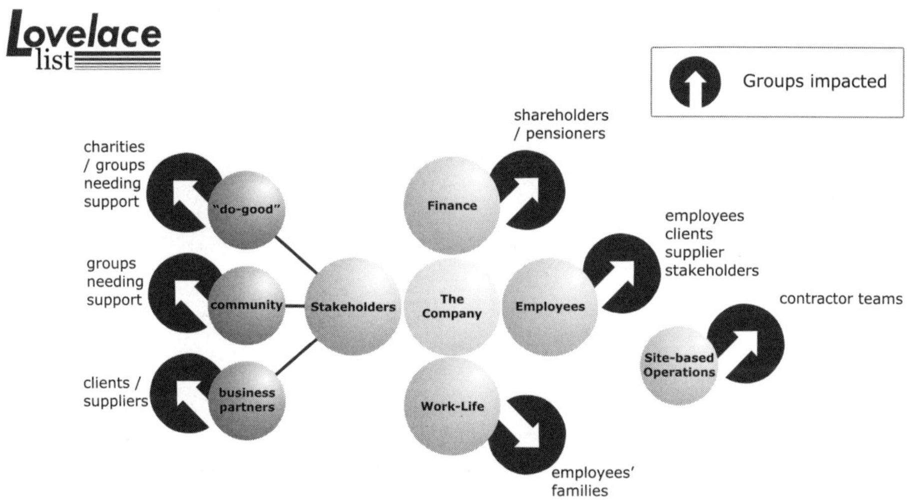

Figure 14.3 The Lovelace List: impacts of imbalance

Since the impacts are essentially on people, it is extremely difficult to estimate the impacts of the outcomes since individuals do not respond consistently to any single event. Similarly, a group response may not reflect the views of every individual in the group.

The impacts map shows the groups that may be affected by imbalance in any one element. These groups can be classified in many ways, see Figure 14.4 for one particular classification, based on their contribution to the company's success, is presented.

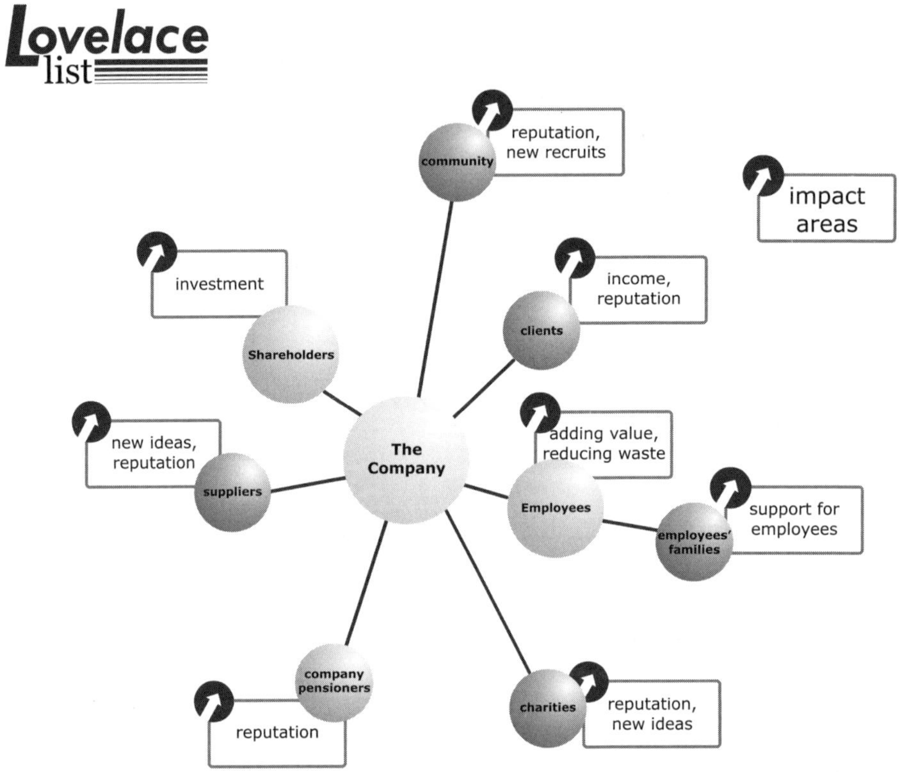

Figure 14.4 The Lovelace List: stakeholder groups and their contributions to company success

The diagram shows a generic view on how each group can contribute to the success of the company. However, a particular stakeholders' contribution for your company may look completely different to what is presented here. What matters is that you should pay attention to the group that can make the most significant contributions to your company's success.

The other side of the balancing act is what elements are best balanced against each other (see Figure 14.5). Again, every situation is different and our suggested matrix shows the levels of balancing effects for each element against every other element.

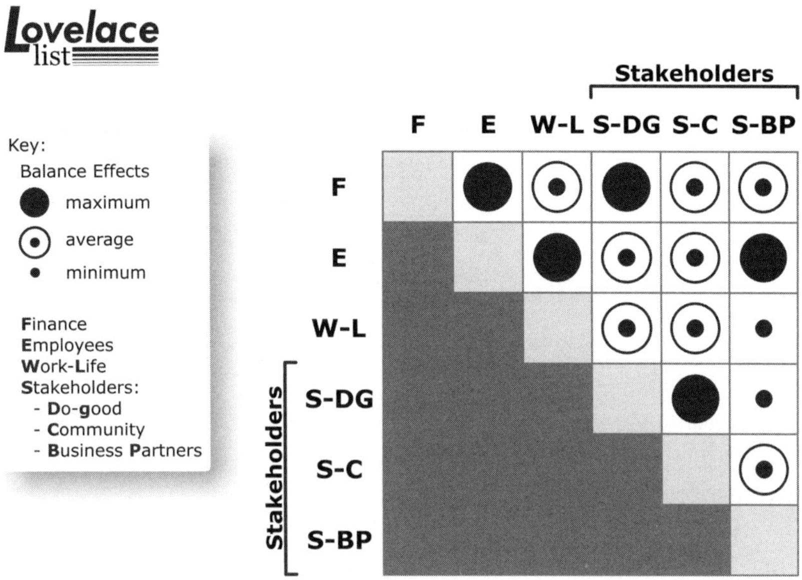

Figure 14.5 The Lovelace List: balancing the elements

The reason for this is effectiveness: every action a company chooses, whether it is financial, social or environmental, needs to be cost effective. Otherwise it creates waste. Therefore the matrix is proposed as a way to maximise the effectiveness of social sustainability activities chosen by a company.

The balancing and needs analysis above, lead to a set of chosen activities that the company believes will contribute towards its social sustainability commitments. Now comes the fun of making it happen.

4. Classic Five Questions Cluster Analysis[1]

The familiar five questions approach of what, how, who, when and why is often used in a casual way but I put into a systematic framework such that the simple tool can achieve very powerful results. Here it is used to work out what to do next.

1 This is described in greater detail in Chapter 15.

Table 14.6 Classic five question analysis

What do you want to do in this area?	The answer provides the outcomes and the targets
How will you do 'that'?	The answer provides the approach or method and the performance measures
Who will work with you?	The answer establishes the team and stakeholders
When will you start?	The answer delivers the schedule, milestones and reporting intervals
Why are you doing it?	The answer clarifies the benefits and defines them in the right context

Answering the five questions also provide an outline for a process improvement project plan.

REVISITING BATNEEC

We return to BATNEEC (see Figure 14.6) because it is a critical aspect of social sustainability. It is a simple check to make sure people are not seeing their investments out of 'green-tinted, designer glasses'. Simply put, it is about getting the best you can afford.

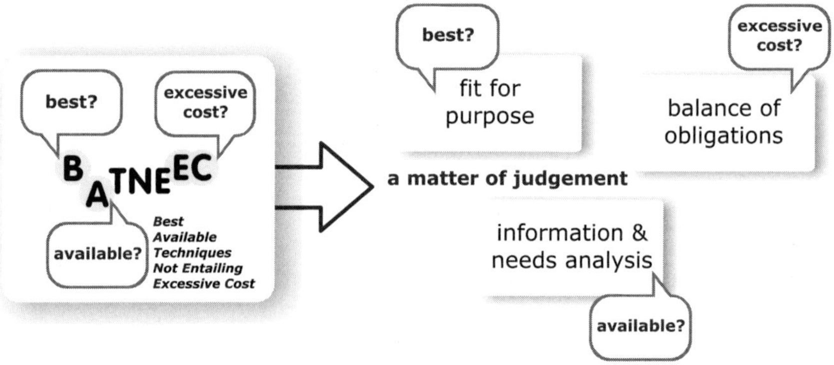

Figure 14.6 BATNEEC

There will be arguments about what is considered 'best', 'available' or 'excessive costs' in BATNEEC, but there are ways to explore this more systematically. The Classic Five Questions Cluster approach above gives you an indication of what you want to do. Other people's best practice shows you what is possible. But you need to make judgements on the BATNEEC components: best; available; and excessive costs.

'Best' – what may be 'best practice' for another company may not be right for yours. Best practice happens because of a unique combination of the right people, right knowledge, right motivation, right budget and right opportunity. These factors are almost impossible to reproduce in another organisation. This is not to say that 'best practice case studies' are a waste of time. They can be seen as motivators as they show what can be done. Furthermore, a cash-rich organisation will develop a set of best practice very different to a talent-rich organisation.

Even in the same organisation, what works well on one project may not translate readily to another. There is a need to assess how each incidence of best practice fits in with your organisation or project site (see Figure 14.7). There may be a need to make changes before you can implement a particular best practice. Therefore, a more pragmatic approach will be to look at what can be adopted and what needs to be adapted. These are then combined with ideas from your organisation to make your own best practice (see Figure 14.8). Companies which understand this concept have no problems in telling you everything about their best practice – they know you cannot simply copy it to make it work like they do.

Assessing what is 'best' means not just taking into account the culture, as we previously discussed, but also the capability of your teams to implement the best practice.

'Available' – each of us only knows a limited number of things (one of the reasons for team working), and the number of best practices we can know of is limited. The 'king of best practice' may be working just around the corner and you would not know about it.[2] The important thing here is the 'scope' or the limits of how far we are prepared to go to in our search for best practice. Without a scope, the perfectionist will keep searching, and procrastinators will keep waiting – and nothing will happen. The aim of all this is to do something, so it is crucial to set the perimeter of how far your search of 'best practice' will go. Do not worry about not knowing every bit of best practice, if you have a commitment to continuous improvement, you will get there and what's more, you are likely to be making best practice yourself.

2 See Jon de Souza's comment earlier in this chapter.

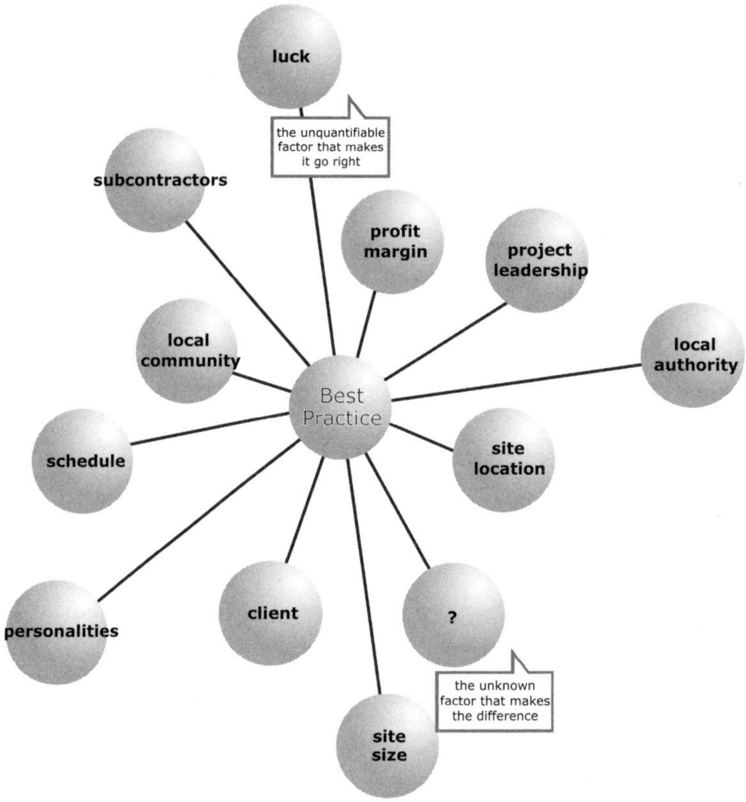

Figure 14.7 How to create best practice: the ingredients

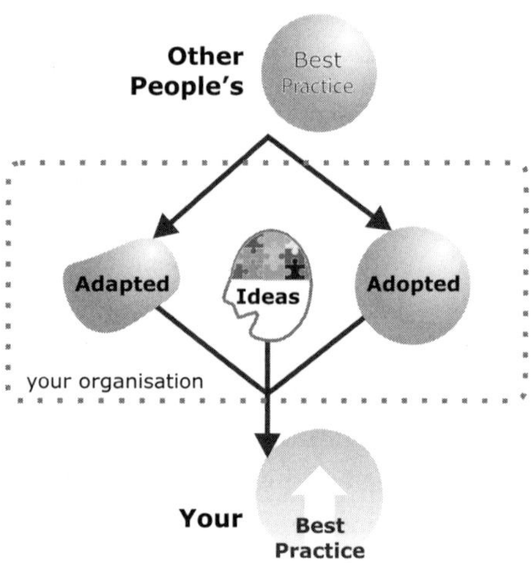

Figure 14.8 The evolution of best practice

'Excessive cost' – this is often the deal-breaker for many people. Since we all have different circumstances, our concept of 'excessive cost' will also differ. There is no getting away from it: if you cannot afford it, do not attempt it. Your banker will not be sympathetic if you ask to reschedule your debt payments because all your money has been spent on doing good.

This creates a dilemma of sorts: do we do only what the finance people say is affordable? (We know what *they* will say ...) Similarly the finance people may think the specialists have absolutely no idea how much things really cost.

We come back to this book's focal point: waste reduction. We can use the money we are currently losing to waste to pay for sustainability improvements. While this sounds good, there is always a fear that any savings from an individual project may just vanish into the corporate accounts with the other profit streams.[3] There is, however, an elegant way of doing this and it is Gain-Share (more details in Chapter 16).

Allocating part of the cost savings from waste reduction to fund social sustainability activities results in a win-win situation for everyone with the following several benefits, including:

- motivation at work to reduce waste;
- clear indication of progress towards the goals;
- openness in linking increased productivity against benefits; and
- opportunities for the company to offer matched funding thus enhancing its reputation.

After examining the available and excessive costs elements of BATNEEC, we come to '**Techniques**'. When BATNEEC was first developed, this was originally described as 'Technologies', but then the management-types got their hands on it and technologies was changed to techniques. Although some engineers were unhappy with the change, the choice of 'techniques' actually makes more sense.

3 The UK government faces a similar problem: should energy taxes be only used to improve energy efficiency? The political jargon for this is called hypothecation. The reality is not so simple and hypothecation is not always easily achieved.

Technology is often represented simplistically by a piece of equipment – a physical asset (basically something you can slap a coat of paint on!). This means that anyone else who can afford it can also get it and share the same technological platform as you. Bang! There goes the technological competitive edge you spent so much to acquire. At this stage, you have two choices, either to continue with the escalation or give up the technological leadership. Both are costly: the former in financial terms; and the latter in reputation terms.

Techniques, on the other hand, is about how you do things. This is not just a 'Methods Statement' but also encapsulates the makeup of your best practice: the corporate ethos and your Core Intent statement. Your decision to install changing rooms and showers for people who bicycle to work expresses not just your ability to afford 'power showers', but your conviction that employees should be encouraged to select alternative transport and also that you express that conviction through the changing room and shower facility. Similarly, having a Gain-Share system is not just about dishing out rewards, but a declaration of intent to share out productivity gains in an open and agreed manner. Clearly the techniques themselves are readily copied by your competitors[4] but how you implement the techniques – your policies coupled with your activities, are not readily copied by your competitors because they are unique to your company and represent a specific set of behaviours.

Figure 14.9 is a simple way of carrying out a BATNEEC analysis: adding value (or not), can you afford it (or not). The hard part is to assess whether it adds value and/or whether you can afford it. This book helps you with the former and your bank (or accountant) ought to be able to help you with the latter!

5. Implementation

Implementing social sustainability activities is not different to implementing any new corporate measure. Nevertheless, since the activities will be about people, a little extra marketing will not be amiss.

Basic marketing tools, such as the Marketing Mix shown in Table 14.7 (also see CD), can be used to help design a marketing campaign to announce the social sustainability activity.

4 Techniques are readily learnt by either reading a book, attending a seminar or (of course) hiring a consultant.

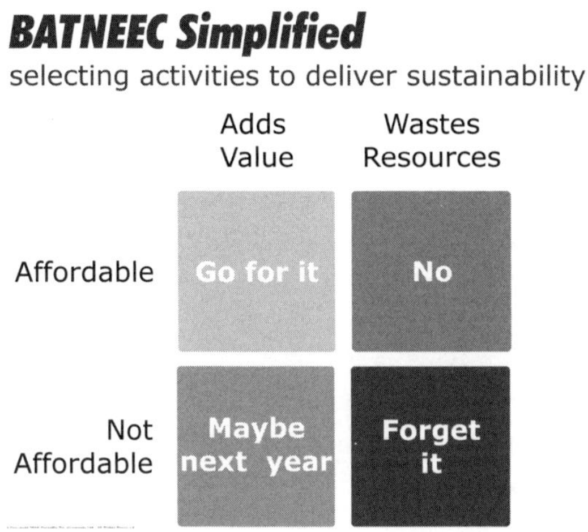

Figure 14.9 BATNEEK simplified

Table 14.7 The Marketing Mix

Product	What is the 'offer'? Do we know what we are buying? What does 'it' do? What will 'it' do for us? How does the 'offer' add value? This is where defining the benefits in context to the people in the company becomes critical
Place	Where does this activity take place? Is it easily accessible to all those who will benefit from it? Do you need to be a rocket scientist to understand it or take part? Do you need to be working at HQ to take part? This is about convenience for the people involved
Price	What is the cost for the employees to making the most of this activity? The costs we are describing include: behaviour switching (see Chapter 16); actual cost (if people need to buy coordinated biking outfits, for example). What is the cost to the company in terms of lost opportunity? For example, if we spent the same amount of money on sales, how much more sales would we have generated? This is about both the easiness of behaviour change and the physical price tag
Promotion	How will we tell our people this is happening? Do we need to provide any other support? (Like bike safety lessons, for example, or do a deal with the local bike store for cheap kit.) More importantly, the promotion message must make sense to the audience. This is about effective communications

And of course, since we have done all this work, we might as well get some good publicity out of it. This is the stage to get the press releases sorted as well.

Finally, do not forget to implement some form of Gain-Share (see Chapter 16), you need to pay for all of this somehow!

6. Measure Progress

Managing social sustainability is managing a dynamic situation, the needs of your employees and other stakeholders may change and you need to monitor these changes to ensure that your investment in social sustainability remains effective and good value for money. A simple approach will be to communicate and survey the relevant stakeholders regularly, a cheaper approach can mean having a steering group representing the relevant stakeholders.[5]

From the Classic Five Questions Cluster process, you would have established some targets and if these are not being reached, then it is time for reviews. Although some programmes can go on and on, others may need monitoring: if one activity is receiving a low take-up, or your ethical investment is being outshone by another ethical investment fund, then it is time for change.

7. Communications

Once you have decided that changes are needed then this information should be communicated to the employees and other relevant stakeholders. One of the basic elements of social sustainability is openness in communications!

Consultation on changes that affect your employees and stakeholders is often useful to gain their input and buy-in towards the changes. Again, you may wish to rank the input from the various stakeholder groups differently in accordance to your corporate needs. As always, do not forget that each group, whether internal or external, has its own agenda and it may not be totally compatible with the corporate agenda. What is more, since most people probably do not know the corporate budget for these activities, it is quite likely that you will receive suggestions for expensive, sensible, practical or pointless activities. Listen, but do not get too side-tracked from your objectives.

A lot of the Lovelace List involves dealing with stakeholders (see Figure 14.10). However, there are always more stakeholders to be found if you are that way inclined. It is therefore time for a reality check since you will never have sufficient resources to deal with all of them. So do you buy them lunch or do

5 Regular communications with all stakeholders and conducting surveys with statistical validity are costly, having the same group meeting every three months to debate (and eat lunch) is much cheaper. However, there is a danger that a steering group can become self-serving and miss the point of communicating with the broader stakeholders.

you adopt their suggestions? Figure 14.10 is a simple but pragmatic way to map and classify your stakeholders. You just need to establish the relevant criteria for influence and impact, and then manage them as your resources allow. A regular review on what matters for your project will give you flexibility in assessing the influence and impact areas.

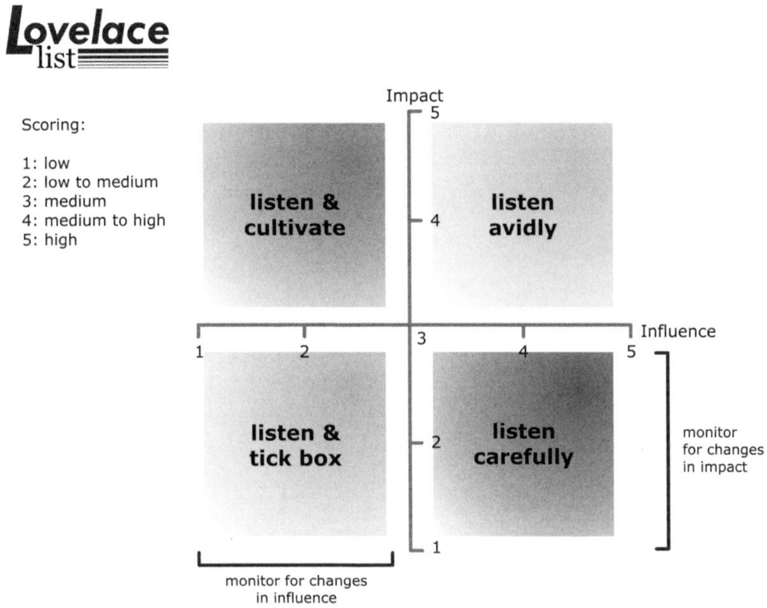

Figure 14.10 The Lovelace List: mapping the stakeholder groups

Integrating the Lovelace List With Business Strategy

The Lovelace List provides a ready checklist for your social sustainability performance and planning. It is also a balanced scorecard to help you ensure an integrated portfolio of activities. However, integration with the rest of the business is often a difficulty when it comes to social sustainability. We feel, however, that many of the analyses and techniques described in this part of the book can be used to identify how social sustainability ties in with business processes.

15

Improving The Way We Work

Doing Sustainability by Adapting the Lean Improvement Approach

When you have an agreed corporate agenda and an understanding of the current business status, there is sufficient information to start looking at how the sustainability performance of existing business processes can be improved through reduce waste and costs through engaging people. The Lean improvement approach is adapted to become the sustainability improvement process (see Figure 15.1).

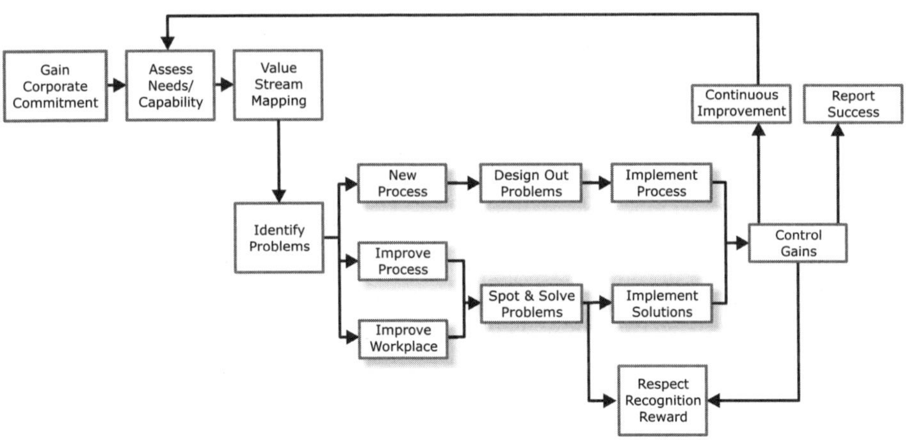

Figure 15.1 Process improvement map

Lean has similar objectives to sustainability: reduce waste; reduce costs; and engaging people, as discussed in Chapters 9 and 11. The sustainability improvement process (see Figure 15.2) has three main streams of activities, these are:

- the new processes stream
- the process improvement stream, and
- the workplace effectiveness stream.

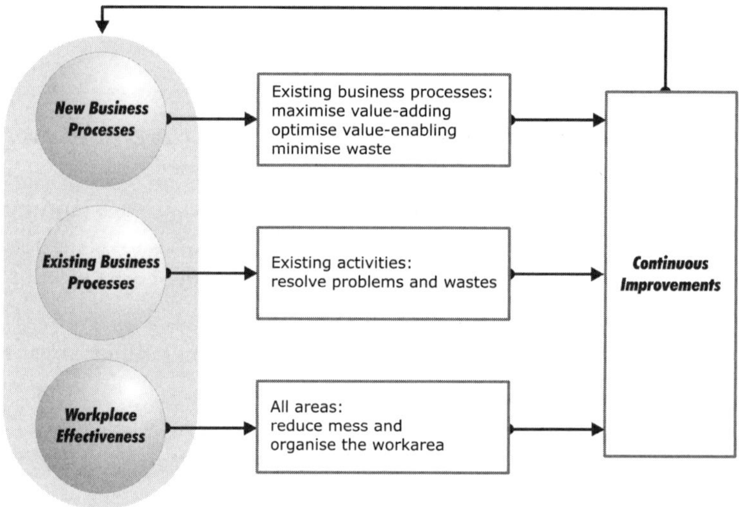

Figure 15.2 Improvement options

These three linked streams are the main areas of what we can do in sustainability through improving our current ways of working. A simple (and low cost) way to approach this is to see whether you can improve the working environment by better design, if that is not possible, then try process improvement, and if that is not going to work, then design a new process (see Figure 15.3).

Other important activities include gaining top-level support; developing an understanding of what your clients want and what you can deliver; and the necessary training make the improvements happen.

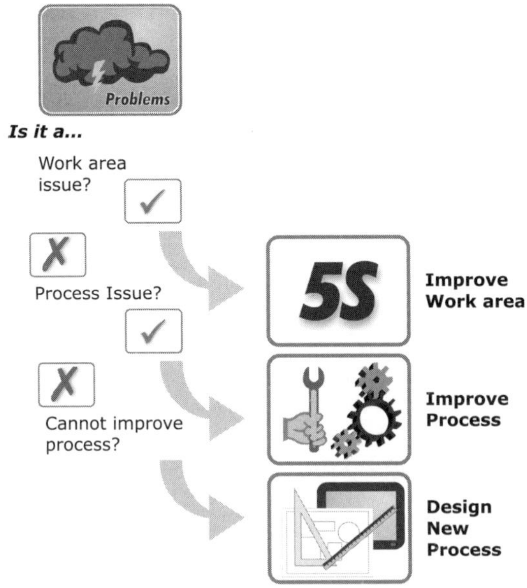

Figure 15.3 Choosing the most appropriate action

Establish Project Level Commitment

This is not a corporate level review like responding to strategic client needs (see Chapter 11), but a working level approach through applying Lean and Six Sigma techniques to deliver sustainability in construction. The project level commitment is probably more difficult to establish than a corporate level commitment. The senior executives from the client, main contractor and subcontractors may meet and readily agree the importance of sustainability to their respective companies and to society. They will also probably agree that much of the implementation should be carried out at the project level supported by a well-worded corporate mission statement. (If you are cynical, you may see this as corporate buck passing!)

But the project director on site has to juggle the realities of delivery, arbitrate between different subcontractors, deal with neighbours and officials from various agencies making their demands (see Figure 15.4). It is no wonder that even with the best will in the world, the resulting commitment can be a bit wobbly at times. The role of the sustainability specialist is really to make that commitment as easy as possible to keep. This means coming up with solutions and possibly compromises to help the project director to keep to the commitment.

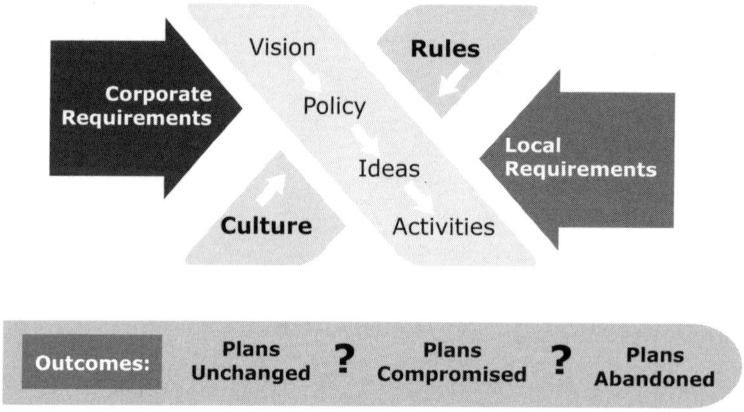

Figure 15.4 External forces on doing sustainability on site

The various factors or constraints that may pull a sustainability commitment off course can include:

Table 15.1 Constraints on sustainability commitment

Vision	To interpret the client's or corporate vision, set the scene and define the way forward
	Potential compromise: sometimes delivering the impractical but well-meaning vision is not easy
Policy	To make concrete the vision and create a framework to support positive actions
	Potential compromise: to balance what is achievable with what is desirable
Ideas	To encourage sustainability innovations and to always do better
	Potential compromise: innate conservatism in our industry may prevent the more innovative ideas, therefore may need to repackage innovation as something less radical!
Activities	To support, to motivate and to celebrate success
	Potential compromise: success may be limited by the skills level of the workforce and a not invented here attitude in the industry
Rules	To control against legislative/policy non-compliance
	Potential compromise: rules must be followed, even if they may be unreasonable and lead to unintended outcomes/tick-box mentality
Culture	To create and sustain the desired business practices
	Potential compromise: balance various parameters to suit a majority
Corporate Needs	To carry out the corporate agenda
	Potential compromise: may not fit local conditions
Local Needs	To meet local (social, political, environmental) demands
	Potential compromise: may need to balance against corporate requirements or client vision

Table 15.1 is not exhaustive but serves mainly as an example, it is likely that you may have other items on your list.

The sustainability specialist can use a range of tools to deal with these situations. However, since it resembles a risk management situation with potential problems, a Six Sigma tool known as FMEA – Failure Modes and Effects Analysis, can be used to provide a quick assessment.

- An FMEA starts with a list of what can go wrong. Here it can be used to list the constraints.
- For each constraint, estimate what are the possible impacts (**I**). Score the impacts, for example, with 1 for low impact and 5 for high impact.
- What is the probability of this particular situation happening? Score 1 for no chance and 5 for 100 per cent certain to happen (**P**).
- What is the likelihood of the situation happening and you did not notice? How hard is it to detect? (**D**). Score: 1 for easy and 5 for very hard.
- Multiply the three scores (**I** × **P** × **D**) and rank the situations for their damage potential.
- Design countermeasures for each situation and allocate the right person(s) to be responsible (probably yourself).

The FMEA is quite methodical yet simple and can provide a workable risk management system for your sustainability programme. As usual, further details are on the CD.

Identify Client Needs and Company Capability

The analysis above showed that just answering 'yes' to a client's requirements is, no longer enough to ensure sustainability delivery on site. But even before we can deal with all the constraints listed above, we need to understand what the client is really asking for:

- understanding what the client is saying;
- guessing at what they really wanted to say;
- having a clear picture of what our company can delivery; and
- guessing (or assessing) whether this is a real capability or an over-optimistic estimate by some colleagues.

Then there is the anticipation of what may happen in the future during the course of the project: new legislation; labour shortages; industry trends; interest rates changes; global financial melt down (or sluggish recovery); and so on.

Client needs and our corporate capability are linked and they are fundamental to any business relation: if we do not have the capabilities, we cannot deliver value to our clients.[1]

We need to know:

- What is required?
- How will we respond?

And then:

- What are our capabilities?
- How will we deploy these?

This part of the sustainability improvement programme is about the establishing the baselines: the demands on us and the values we can deliver. At this point, the Six Sigma discipline of relying on hard data comes into play: we need to audit our capabilities as well as survey our client's viewpoints. Unfortunately, both of these are likely to be constantly changing, so the job is not an easy one. However, we need to measure our capabilities and our clients' requirements, otherwise we cannot hope to manage either effectively.

1 You do not want to be in the position of a senior manager of a major organisation who told me: 'we have pockets of excellence in our company, but we just don't know where they are'.

The voice of the customer (VoC) is an approach used to describe the client's needs. It provides us with the necessary information on what needs to be done. These can be found by survey questionnaires as well as interviews. A simple technique to sort out all the different voices (or needs) is the Affinity Diagram where similar items are grouped, the duplicates eliminated and common themes emerge.

Of course, when we listen to the VoC, there is a risk of information overload, this is especially when we are dealing with lots of voices from multiple stakeholders. In developing a basic product, a brick, for example, the voices are generally clear, but for a sustainability activity, this gets complicated. Conflicting voices on the financial issues, social issues and environmental issues can come from within the company, and then there are the external stakeholders.

The situation with a broad group of stakeholders can be likened to a jigsaw puzzle (see Figure 15.5), there are many pieces but not all are essential for you to get a sense of the picture. We do not have the resources to check and follow through all the voices. Using the affinity diagram technique may mean we lose a bit of detail, but we get an overall feel for the situation (and stay within budget).

Figure 15.5 The sustainability jigsaw

Our responses can also be improved by divining where the client's requirements fit on the Critical to Sustainability Analysis value map. This knowledge will allow us to make a more effective response because we will not be over delivering, one of the Eight Wastes from the Hidden Construction Site!

Value Stream Mapping (VSM)

The VSM analysis for process improvement is the same for strategic business process or local processes, as described in Figure 15.6. The difference is that at the local or working level, it is about the local construction processes. The working level VSM is generally about walking around the site physically and following a procedure or method.

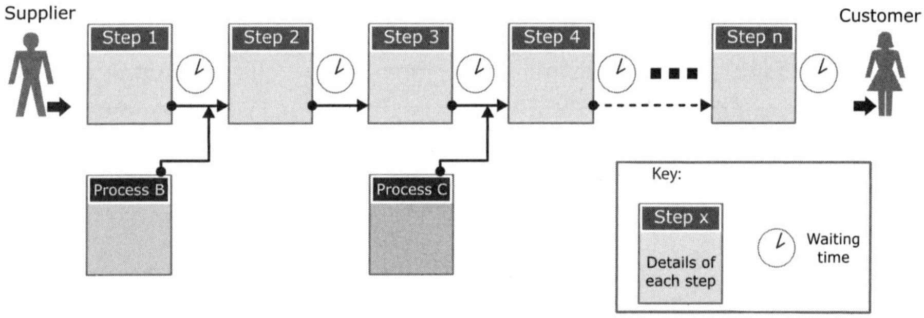

Figure 15.6 Value stream mapping

The working level VSM can be time-consuming because it requires someone (you probably) to 'walk' the process and record every step in some detail through the value-adding elements of the process on a construction project. The alternative is to sit at your desk and guess the process. This less strenuous approach is likely to deliver the sort of outcome you expect from guesswork.

The initial VSM analysis produces a map that is the 'current state'. This provides the necessary facts and data about the entire process listing every step: what is happening; who is doing what; what are the values; where are they added; what is the waste; and where do they occur. This describes the way work is done currently and we can see which activity needs improving, fine tuning or replacing. This analysis usually results in a long list of areas we can improve.

You map the process by logging the details of every step or unit activity, and these can include virtually any parameter you choose. Choosing wisely can result in useful information rather than an unmanageable mass of data. Measuring the effectiveness of each step can be carried out through the Overall Effectiveness (OE) methodology. (OE is part of Total Productive Maintenance). The OE of every step of the process can be assessed from three factors: the availability of resources; the performance of the resource; and the quality of the output, as shown in Table 15.2.

Table 15.2 Overall effectiveness methology.

Factor	Indicators
Availability	Physical availability of people, plant, information and materials; actual amount of time worked is carried out; duplication; over-ordering and clutter
Availability rate = actual resource availability for deployment / maximum designed resource availability (%)	
Performance	On time or late; degraded materials and information (passed their used-by date); people with inadequate or wrong skills: wrong tools
Performance rate = actual progress / designed progress (%)	
Quality	Defects; rework; rejects; typos; wrong information and not meeting specifications
Quality rate = number of defects / total number of opportunities (%)	
OE = **A**vailability rate × **P**erformance rate × **Q**uality rate (%)	

All of these factors can provide an indication of how the process is doing. The aim of the value stream mapping process is to decide when the value-adding happens (ideally, every step should add value) and where the waste happens.

When carrying out the mapping, you should remember the following 'to do' list:

- Communicate your intention to the relevant people before your mapping visit.
- Get the agreement of the local team leader or supervisor before asking questions.
- Introduce yourself clearly to the team.

- Explain what you are doing and answer questions openly and honestly.
- Remember the people working there are the experts in that particular area of work.
- Ask them about the problems and remind them that there is no blame.
- Remember to thank the people involved (and do not offer any suggestions on how they should do their work or their improvement ideas at this point!).
- Once you completed your survey and prepared a Future State, go back to the team and check whether the 'new' version will deliver the goods.

After you gathered the data, you can assess where the value is in each step and where the wastes are. The aim is usually to focus on the areas with poor performance, but you may wish to show some recognition of the value-adding steps as well. There may be some lessons that can be adapted for the less efficient steps.

Alongside the VSM, the SIPOC (see Figure 15.7) analysis is useful for analysing the individual process step. SIPOC is a Six Sigma analysis that looks at the relationship between a single process step and its immediate upstream and downstream activities. SIPOC[2] stands for **S**uppliers **I**nputs **P**rocess **O**utputs **C**ustomers. The separation of the people (suppliers/customers) with the resource (inputs/outputs) allows a greater degree of analysis since issues, such as materials storage and retrieval, can be assessed on its own merit. You may find that defective materials to be a cause of waste, but is the defect from the upstream supplier or from our storage of materials? Similarly, when our downstream customers complain about our delivery, is it because our process is slow or is it the dispatch of our output that is holding up the process?

SIPOC is almost a miniature VSM for one particular step in the value stream. The usefulness of SIPOC is that it does not isolate the individual step (the process) from the entire value stream by including the impacts of the

2 I think it is pronounced: 'Si-poc', 'See-poc' or 'Cy-poc'.

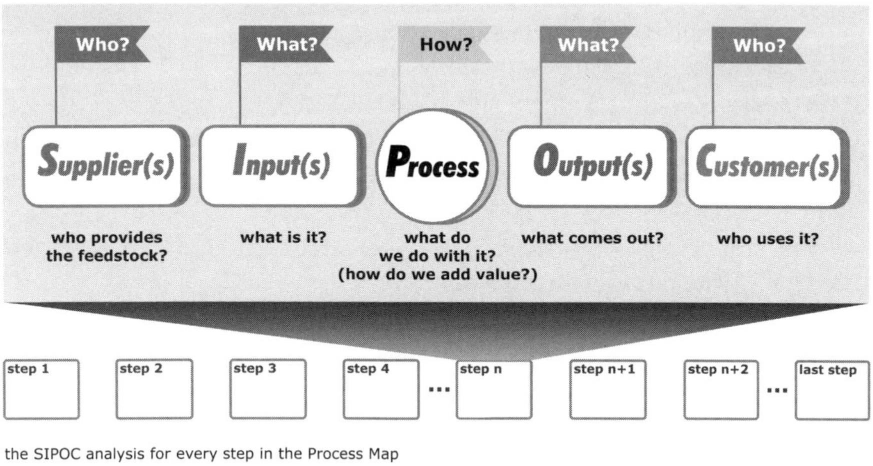

Figure 15.7 The SIPOC analysis

upstream (inputs) and downstream (outputs) activities. My suggestions for using SIPOC are:

- steps that are complex;
- steps that contained a lot of waste; and
- the handovers (materials, semi-finished parts, information) between the formal process steps.

One other aspect of the working level VSM is the actual mapping of the physical movements of materials and people in the work surroundings. These transportation diagrams are also called workflow diagrams and 'spaghetti diagrams'. They show the human factors of workplace design. If we lay out our workplace in an inefficient way, people not only have to do a lot more moving around, it also wastes time, increases the risk of accidental damage to the components and perhaps even accidents to ourselves.

Value stream mapping and the SIPOC process are described in more detail on the CD.

After the value stream mapping, three parallel streams of activities described in Figure 15.1 and Figure 15.3 can be developed that describe the way forward, they are:

1. workplace effectiveness stream;

2. process improvement steam; and

3. new process stream (working level).

Table 15.3 The parallel streams of activities

	Effort	Costs
Workplace effectiveness	Low	Low
Process improvement	Medium	Medium
New process	High	High

These three areas form the main part of sustainability improvements at the project level.

1. The Workplace Effectiveness Stream

People work better in more pleasant and safer surroundings. If the workplace is not well organised, you may have to climb over mounds of rubbish to get to your workstation or twist your body around in order to do a job, or having to move things back and forth. None of these contribute to greater effectiveness, but instead may add to the risks at work and make you a lot more stressed and tired. This is a first step to more accidents, errors and mistakes. The same goes for people having to work in filth and mess. Since a construction site will never be like a serviced office, we cannot expect the same level of cleanliness or tidiness. However, we can and should expect a workplace that allows us to deliver the best work possible.

Toyota Motors recognised this in the late 1950s when it realised that to be competitive with the American car makers, it had to provide a platform where innovations and more effective working can take place. Similarly, for our teams to deliver maximum value, we need to provide a working environment that takes away unnecessary hassles and hazards. The Toyota solution is called 5S, from five Japanese words starting with 'S' sounds.[3] (5S is explored in more detail on the CD).

Table 15.4 The five 'S's

Sort (*Seri*)	Get rid of clutter, duplicates and unwanted materials (for reuse, if possible)
Set (*Seiton*)	'Everything in its place and a place for everything', a less stylish way of saying this is: make sure you store things properly so that you can find them later
Shine (*Seiso*)	A clean and clear workplace is safer, more pleasant and a more effective workplace
Standardise (*Seiketsu*)	Simplify the processes and methods and stick to them, therefore reducing errors and variations
Sustain (*Shitsuki*)*	Keep up the effort, review progress and carry out continuous improvement

* This is not pronounced as you may think – the first syllable ends with the first i: *Shi – Tsuki*!

Although simple sounding, the 5S workplace effectiveness programme is crucial in almost any Lean or sustainability improvement programme. Although the first 'sort' exercise will mean an initially heavier skip or a bigger stack for reuse, future skips will be lighter as over ordering is avoided. Doing 'set' will mean labels are securely tagged and people have a better idea where information and materials are, so wastage is reduced. 'Shine' is partly about the social aspects of sustainability and reducing the risks of slip, trips and falls. By keeping the workplace clean and clear, raw materials and semi-finished components are less likely to be damaged. 'Standardise' reduces errors and people taking short cuts (well, you are unlikely to eliminate these completely), increasing quality and cut down materials waste. 'Sustain' is perhaps most fuzzy since it does not involve skips and bins or stacking blocks neatly. But it is about the determination of the entire team in keeping the workplace effective.

The benefits of 5S, a well-organised workplace, cover all three elements of sustainability. It has been sold by consultants as a money saver, it should be sold to health and safety managers as a tool to reduce accidents and hazards. It

3 As most people cannot pronounce (or remember) the five words in Japanese, various sets of English translations are available, but since these are invented by consultants, they are often incompatible and serves to confuse at times. My version is from Productivity Press, which publishes books on Lean manufacturing.

will, however, upset the government and waste management companies since you will have less waste, thence paying less for disposal and Landfill Tax!

The other benefit of 5S is that it results in a clearer, neater and tidier workplace. Since the Health and Safety Executive reckoned that nearly 50 per cent of accidents on side comes from poor housekeeping and messy organisation, 5S is also a very useful implementation to reduce hazards and accidents on site[4] and can be part of your social sustainability programme too.

2. The Process Improvement Stream

Although designing a new process can result in major gains in productivity, such a major revamp takes time and while that is taking place, you can still do something positive by carrying out minor improvements – lots of them. The process improvement stream takes the problems listed in the value stream mapping and uses various tools to deal with them. These minor improvements can lead to a significant competitive advantage because first, small savings add up quickly and second, the large number of small improvements mean that your competitors are not likely to be able to copy them all. Third, the discipline from implementing a series of improvement builds up an ethos for continuous improvement. Finally, carrying out improvements continuously means we become more experienced in solving problems, thus enhancing our capability. Problem solving results in innovations and provides incremental gains in effectiveness. Since our rivals are also improving the way they work, a continuous improvement ethos (or edict, depends on how your company sees management) will push us forward and help reinforce any competitive advantage.

Of course, having innovations is useless if the good practices are not maintained. There is nothing that is more disappointing than an improvement programme being eroded and the efforts and investments lost. Controlling the gains is crucial for the sustained success of any sustainability improvement programme. But it is also very difficult to carry out in the construction industry because of the nature of subcontracting in our industry. Unless your team remains constant and this includes the subcontractors, there is a pretty good chance that some of the gains will be eroded.

4 The Health and Safety Executive reported this in HSE Causal Factors in Construction Accidents – Report 156, 2003.

The framework approach used in the construction industry is one way to ensure good practices and improvements are not lost from the partnership. However, the recession of 2008–09 has caused some companies to save money by abandoning the partnership framework approach and instead go for the lowest bidder approach (as reported in *Building Magazine*, issue 25, 2009).

An alternative is to create a set of 'good' working methods that are up to date and relevant. By sticking rigorously to these working methods, you can overcome staff turnover issues with disciplined working practices. This way, the battle does not have to be fought every time, because any gains in efficiency can be put into an updated set of Method Statements, Tool Box Talks, guidelines and procedures that managers and supervisors can use to help all staff to become increasingly comfortable with sustainable ways of doing things.

3. New Process Stream – Pull Instead of Push

Although virtually all processes can be improved somewhere (and somehow), there comes a point when it may be more efficient to redesign the entire process from scratch.[5] There are many ways to do this but one Lean approach is to design it backwards.[6] Starting with the finished product or service and ask: what do I have to do to get here? And going back step-by-step by asking the same question. This is an inherently logical way of designing a process to meet both client and sustainability requirements.

By starting with the outcome itself, the focus is on delivering value to the client. So client needs drive the entire process. In Lean, this is called 'pull', the opposite of which is 'push' (see Figure 15.8). Push is where the company pushes a process to the client, with all the wastes, bureaucracy and other problems because 'this is how we always do things here'.

A 'pull' design will have only the necessary activities in the process because nobody will design any of the 'Eight Wastes' into a process or add bureaucracy to a process. The end result is often a shorter process and one that uses fewer resources. By cutting out the unnecessary steps, all three elements

5 The jargon for this is sometimes known as re-engineering.

6 There is a related but very different Lean process tool called the Last Planner, which was developed by the Lean Construction Institute. Check it out at their website (www.leanconstruction.org/).

of sustainability are satisfied: you have fewer steps, so the costs (both direct costs and overheads) go down, people are happier, and if you are using fewer resources, it is likely that your environmental impacts will also reduce too.

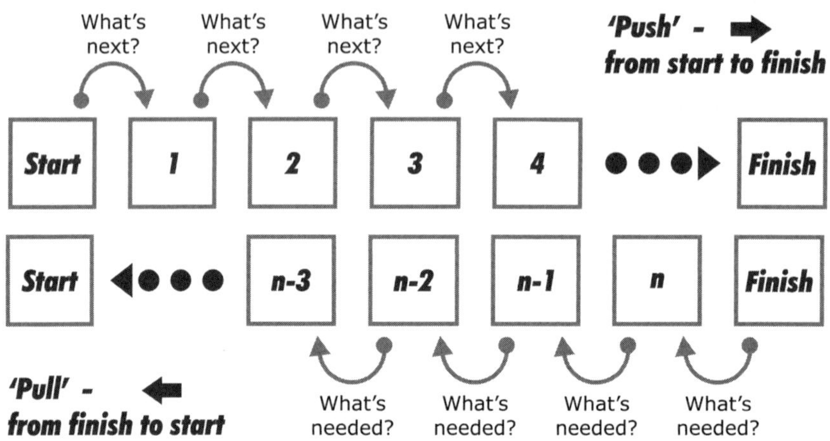

Figure 15.8 Push and pull processes

Designing Out Potential Problems

Once you have the bare bones new process, you will need to check it for potential problems, the FMEA approach described earlier is also a good way to examine this. Another way is by a Lean concept using a Toyota idea called *poka yoke*,[7] or fool proofing. This means designing processes, products and systems such that essentially, you cannot mess up unless you try very hard. The assumption is that most people are too lazy to try *that* hard to mess up. A good example is the 110v electrical plugs used on site (see Figure 15.9). You can only plug it in one way. Designing out problems requires considerable planning and designing expertise as well as real experience. In construction a possible deployment of *poka yoke* is to lay out working areas and also in the packaging of materials. Redesigning all the method statements using *poka yoke* is a good idea in principle, but will be quite time-consuming.

7 I have no real idea how this is pronounced, my assumption is that you pronounce every syllable. The original version was called *baka yoke* (idiot proofing), but probably changed to poka yoke because of political correctness.

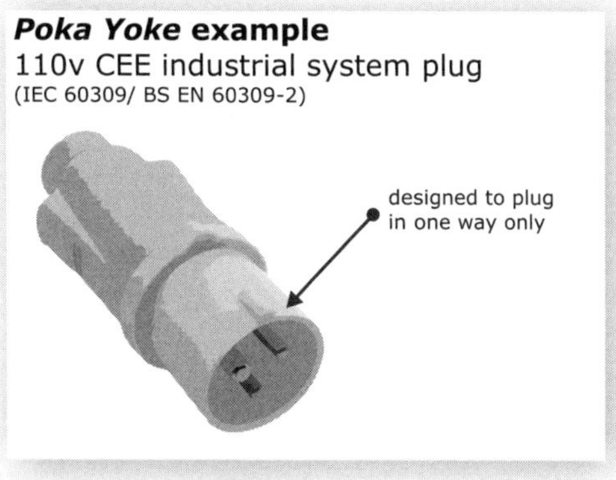

Figure 15.9 A Poka Yoke example

Making Improvements Happen

The various analyses on the processes and the workplace have identified problem areas where improvements can be introduced to reduce waste and inefficiencies. The next stage is about solving the problems and delivering the improvements. However, before that can take place, you need to decide how you want to proceed with the actual improvement activity.

There are a large number of ways to bring about improvements, do you:

- Go to the boss for a solution? The boss was hired to solve problems, no?
- Guess at what may be the cause and then solve the problem?
- Apply first principles and come up with a solution?
- Believe that your solutions will work if you tell people what to do?
- Hire a team of consultants and let them come up with the solutions? (Yes!)
- Tell the subcontractors it is part of their job to come up with solutions?

Who Solves the Problems?

There is a traditional management approach where it is believed that leaders, managers, bosses have some mystical ability to solve problems. In these environments, problem solving is the realm of the bosses. Basically, the boss decides what the problem is and proceeds to solve it and life goes on. However, this approach has an inherent blind side – the boss cannot be everywhere and cannot know everything.[8] In other organisations, problem definition and solution often default to the loudest voice, whether or not the necessary capabilities are present!

Problem solving is often an automatic process in our minds, but a more systematic analysis shows that it is made up of several attributes in solving problems. This also highlights the problems about the boss being the know-it-all.

Since sustainability belongs to everyone, it is only sensible if everyone is involved in getting to the solution. You may feel that the solution is blindingly obvious (this happens whether you are solving it in a team or on your own), but one of your colleagues may have thought of something you did not and that can change the whole analysis. The other argument is that each manager has one set of eyes and can only see one thing at a time. The team possesses more sets of eyes and can spot more things than a manager at any one time. This is not to say that a leader's contributions are not valuable but it should not be all prevailing (*'this is the solution because I say so'*). A manager contributes the information that the team may not necessarily have, such as a bigger picture comprising further upstream or downstream activities. Having said that, there must also be a careful balance in the roles and functions as the manager will need to be accountable for the team's decisions.

In our industry, a common approach is to 'tell people what to do' – I have heard this from conservative construction managers to with-it HR managers. It is the same, we are telling people what to do instead of tapping into the ideas and potential innovation of their workforce.

The approach I propose is instead a team-based problem solving involving both the boss and the team. The recommended tools are described in the next section.

8 You know this, so do I and if the boss is honest, s/he will admit it too, but some diplomacy will be useful in dealing with this.

Table 15.5 Attributes to problem solving

Attribute	Boss	The Whole Team
Knowledge	Extensive but may be at a higher level and not necessarily up to date or particularly practical	A working level outlook, less strategic but knows the problems and has to live with the daily work rounds
Experience	Extensive but broad	More focused
Skills	Probably knows more about management techniques	Probably more skilled in running the process
Competence	Variable but perhaps team may have less confidence in the boss	Variable but more trust from other team members

There are a few tips about solving problems as a team.

- The boss does not need to be there for every problem-solving session, but instead has to delegate the problem-solving authority to the team.
- The boss needs to be sure about the team's problem-solving capabilities or provide support to boost skills.
- The boss needs to support and respect the decisions from the team. But also question decisions in a positive way.
- The team needs to respect the boss' decision to delegate and use the authority with responsibility. (Don't embarrass the boss.)
- The team should report back to the boss after any decision that results in changes to the working method or process, or impacts on other teams.
- The team should ask the boss for help or request additional specialist support when required.

Yes, team-based problem solving will take longer, and yes, it may cost more right now. But in the long term, a team that is good in solving problems will deliver much more value. Like most things in life, team debates are not always suitable for all occasions: where health and safety is concerned, common sense and following safety instructions has to come first. Once everyone is safe, the team can then debate the root cause of the emergency!

After using the VSM to gather information on the weak points of the business process and organisational issues of the workplace, it is time to explore the waste problems with the team.

Solving Problems – a World Class Technique

A very simple but effective framework for solving waste problem is a standard Six Sigma tool, DMAIC[9] or **D**efine **M**easure **A**nalyse **I**mprove **C**ontrol. Six Sigma is based on making decisions based on facts and DMAIC demonstrates that approach:

D (define): define the problem and its impact(s).

M (measure): measure the extent of the impact(s).

A (analyse): analyse the problem and arrive at solution(s).

I (improve): improve the situation by implementing the solution.

C (control): control the gains from the improvements.

Unlike the other tools in this book, DMAIC is described more extensively in the book rather than the CD because it is one of the most useful tool in solving just about any problem.

Define

Defining a problem sounds easy, but sometimes it is not. Take, for example, a very messy work area: is the problem the mess or is it a safety hazard, or is it a source of materials waste? The path to solving each may be different as may the solutions. This is where DMAIC's insistence of going to where the action is becomes important.[10] Direct observation with the team involved usually resolves the problem definition. It is also safer than guessing from the office.

9 DMAIC is apparently supposed to be pronounced 'de-mack' or 'de-may-ack', but that is not a strict rule!

10 This is also a Lean/Toyota approach '*gemba genbutsu*' – go to where it is happening, see for yourself whether it is work area or where the situation occurred and get the real facts and data.

Where direct observation is not possible then photo evidence or reliable third-party evidence should be used – but stick to facts not opinions!

The aim of the define stage is to ensure that the team gain a shared view of the problem. It may be helpful for the team to use the What/How Matrix, as in Figure 15.10.

By asking, 'what is the problem?/how did it happen?' rigorously, it is possible to isolate and clarify the problem; or alternatively use the SIPOC analysis described earlier to differentiate the problem among a mass of observations. The What/How Matrix is also used extensively in the House of Sustainability analysis (see Chapter 17).

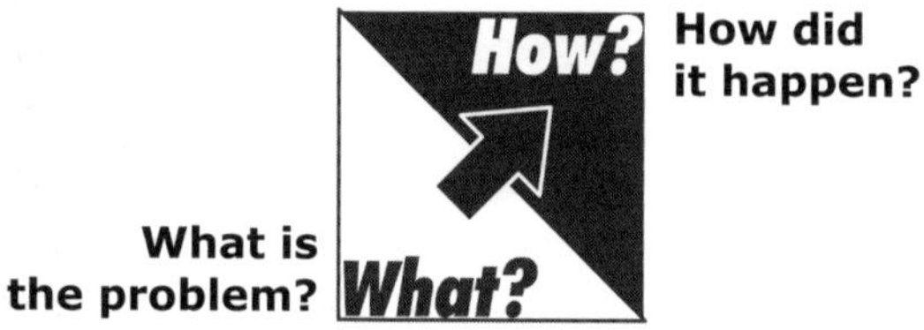

Figure 15.10 The What? How? Process

Measure

We are rarely that lucky to have just one problem to solve: sometimes problems are like buses, they arrive all at once. One of the benefits of using DMAIC is that it can help us prioritise. By *measuring* the impact of the various problems, it is possible to get realistic data of the problems and their impacts.

In terms of assessing a waste problem, there are a number of parameters to choose from. The VSM or 5S often provide the necessary underlying data for any analysis. There are only two rules in measuring the impacts:

1. **Commitment**: try to get agreement about which measure(s) to use for all the problems.

2. **Consistency**: always use the same measures for comparing the same type and batch of problems.

A starting point for a more rigorous analysis can be the Critical to Sustainability Analysis.[11] This takes into account the impacts on the financial, social and environmental elements. You may wish to select other elements to measure the impact. A good one is money or costs. Just about anything (delay, damage, H&S risks, pollution, waste, productivity) can be translated into money terms.[12] When every problem is directly comparable with every other problem, either via the Criticality Analysis or translating the problems into a common term like money, prioritising becomes much easier and individuals and teams are more likely to accept it although they will still moan about it! The prioritising techniques of aligning everything and then select the ones with the most impact is called the Pareto analysis or the 80–20 rule. The idea being that *80 per cent of the impacts comes from 20 per cent of the problems.*

Analyse

The DMAIC tool also acts a framework for deploying other tools, see Figure 15.11.

Again the aim is to make sure the team achieves a shared solution. Several tools from Lean operations and Six Sigma can be deployed in these analysis situations. (See the CD for more details on how to use these tools.)

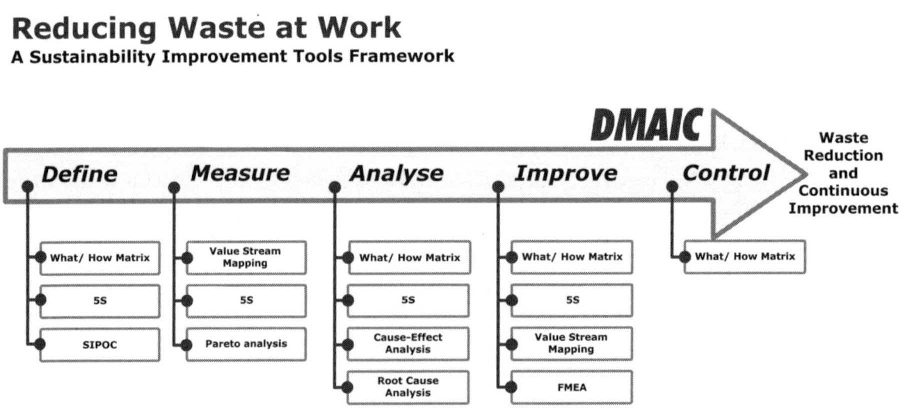

Figure 15.11 The DMAIC framework

11 See also Chapter 1 and Figure 1.3.

12 Once you managed, via some form of voodoo economics (or ask a management accountant nicely) to convert a 'demotivated team outlook' (or whatever) into £, $ or €, you may then change the terminology: 'cost' or 'waste' into 'potential cost savings', not only does this sound more positive, it also gives the team a sense of purpose when they solve the problem.

1. **What/How Matrix**: by asking, 'what happened?/how did it happen?' and then cascading to the next level ('then what happened? And how did *that* happen?'), it is possible to establish the course of observed events as well as gaining information of how it happened at every stage.

2. **Workplace effectiveness**: 5S (**S**ort/**S**et/**S**hine/**S**tandardise/**S**ustain) provides possible clues to identify the areas where the problems are observed. Like the What/How Matrix, the 5S provides observed data for the next analysis.

3. The **Cause-Effect analysis** is also known as the Fishbone analysis or the Ishikawa diagram.[13] This is a team-based tool and its purpose is for the team to agree what are the likely causes of the problem. It is rigorous and somewhat labour intensive, so do not embark on this when solving a problem on your own unless you need to go through a lot of confusing data. This analysis is an effective sieve and you use it to sort out all the possible causes and end up with one that everyone on the team agrees and buys into. One of the joys (?) of the Fishbone is to see stubborn individuals defending their pet causes very loudly (and persistently) to other team members (see CD for more details).

4. When the team has reached an agreement about the most likely cause, the next step is solution generation. A quick and effective tool is the **Root Cause analysis**. This is another Toyota tool that is also called '5-Why'. As its name implies, the question 'why?' is asked successively until the root cause is reached (see Figure 15.12). The Root Cause analysis is structured such that the answer to every 'why?' provides the clue to a solution. The 'because ...' answer describes an event which, if changed, will prevent the problem from returning.

5. The Root Cause analysis is one of the simplest tools yet it has several subtleties.

13 Being a phonetic translation of a Japanese surname, it is sometimes called *Ishigawa* diagrams.

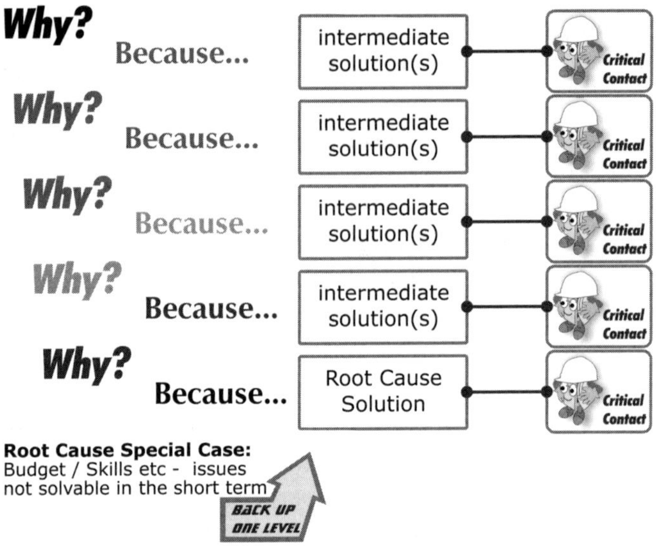

Figure 15.12 The five 'Whys?'

Subtlety #1: Every 'because' offers an intermediate solution. This solution will deal with the current problem now, but will not necessarily prevent the problem from reoccurring.

Subtlety #2: Behind every 'because' is often a person who probably designed the job in the first place. This person is a 'critical contact' and needs to be on board for the change you are proposing. The critical contacts are also likely to resist change since you are going to overturn their original decisions!

Subtlety #3: Sometimes the root cause proves not readily solvable – such as a lack of budget or skills. The way forward is to go *back* up to a level that can deliver an interim solution. The root cause issues are logged and dealt with then the situation allows.

Subtlety #4: The initial question needs to be framed with some care, a broad question may branch out and cannot be solved within five sets of whys. A tightly focused question derived from the 'most likely cause' (Cause-Effect analysis) is more likely to give a useful root cause analysis.

Subtlety #5: The outcome of a Root Cause analysis is a often technical solution. Although this is 'the solution', it is not complete because it is not likely to work in practice until you have convinced the team involved to adopt the solution. Even though the team itself may have come up with the most likely cause and then derived the root cause, there is still the chance of resistance. And if you have to deal with an awkward critical contact, life can get a lot tougher. This moves on from solving problems to managing change and is discussed below.

Subtlety #6: In a complex situation, you may have several streams of activities that lead to the sustainability waste problem. In these cases, the Root Cause analysis branches into the various streams and every stream needs to be resolved to ensure the problem does not reoccur. For example, if your problem is a pallet of water-logged bags of cement, then a purchasing focus may look at over ordering, a store focus may address lack of proper storage space and a working-level focus will ask why did we not cover up the cement pallet.

Despite the subtleties, the Root Cause analysis is often a quick and thorough way of getting to a solution (see Figure 15.12).

Improve

The improvement or implementing the solution is a mix of both the technical solution and any necessary sugar coating you put on to make the change easier. Change is not easy (see Figure 15.13) and it is not always comfortable for the people involved. Just try this simple experiment for yourself – cross your arms normally, now reverse that arm-crossing. Is it comfortable? Can you keep crossing your arms in this new way for the next day or so? This is an extremely simple change, yet even here it is not easy. Think about how hard it is when the change is about something that really matters.[14]

14 You have just been told to cross your arms, but given no reasons why or any benefits explained. The chance that you will make an effort to change is slim, but if I was to offer a benefit (say, for example, free breakfast if you do it) then there is an incentive and your decision towards change may shift.

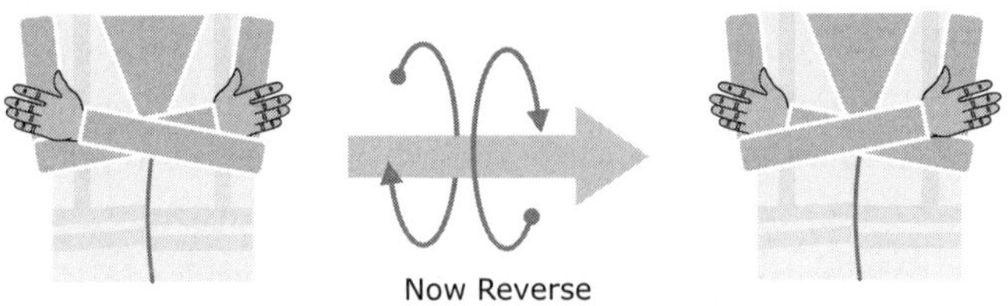

Figure 15.13 Is change easy?

Tool Box Talks, a staple of the construction industry, are used to pass on important safety message relating to the work. However, there are other ways to use the Tool Box Talks. Both the NoWaste programme I developed for Bovis Lend Lease or Project Sus'd for UCATT uses Tool Box Talks to spot and solve waste problems.[15] There is no reason why other companies should not use Tool Box Talks similarly. This is an occasion when the team can discuss how they can contribute towards reducing waste. In this respect the Tool Box Talk[16] becomes a combination of problem solving and training session. It also reinforces the continuous improvement ethos.

The Classic 5 Question Cluster (see Figure 15.14) is another way of using a familiar tool (what, why, who, when and how) to provide a very effective way for a team to use the tool box talk as a way to come up with waste reduction ideas for sustainability improvement.

You may think that just because the team came up with the solutions, they should buy into it readily. After all there is ownership and all the management books tell you that ownership is a prerequisite for accepting the change. Well, let us use the New Year's resolutions as an example.

15 The latest version of the NoWaste tool box talk is known as NoWaste *Espresso*. This version was launched at the big Stratford project during the Sustainability Week in November 2010 – see CD for more details.

16 Several tool box talks of this type are in the CD.

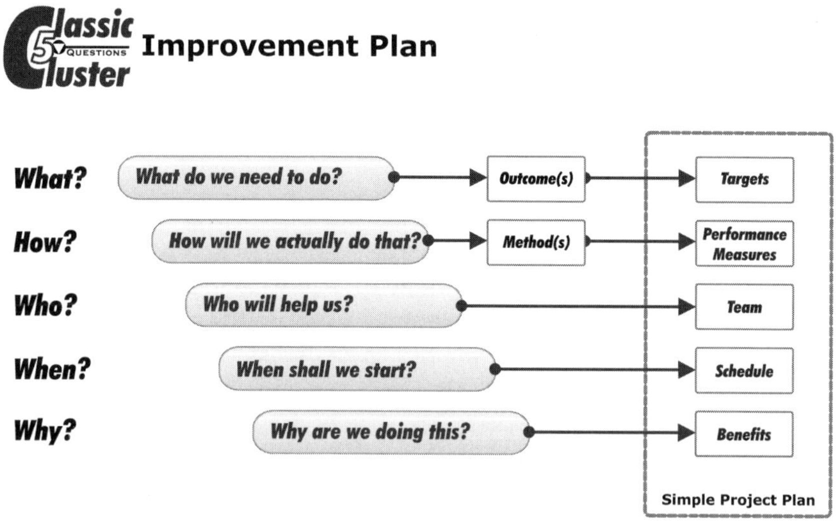

Figure 15.14 The Classic Five Questions Improvement Plan

Table 15.6 New Year resolutions example

Thinking About the Problem and the Changes Needed	
Problem Definition and Measurement	You look at yourself in the mirror and think: I don't look as good as I did and I can look better
Analyse	You look at the articles in the newspapers and found that working out at a gym gets you fit and improves your health
Improve	My New Year's resolution is to get toned up in the gym
You have ownership of the solution and you are willing to go through a tough change. Now comes the 'doing' bit about change	
Improve	Gym membership paid for
Control	A few cold and rainy January mornings and your resolution crumbles
Result: not a lot of change. Owning the solution does not necessarily make you implement and sustain the solution. You also need on-going motivation, support and rewards. A sugar-coated donut may not be the best thing, but it may just be your reward for going to the gym	

However, at work, you need to think about the sugar coating quite carefully. People refuse or reject change for a number of reasons. Our industry is quite traditional in its management practices, in other words, new management practices are slow to gain much take-up.[17] You can imagine a supervisor or operative giving you a look that says variously, 'you are an idiot' or 'we never did it that way with Laing O'Rourke/Balfour Beatty/Carillion/Kier/Costain/ Bovis', or 'this is the most ridiculous thing I've heard in 20/30 years'.

You can probably imagine a response like this from the critical contact (see Root Cause analysis) or the person who has to work in a new way. These people have a stake in your solution since he/she probably came up with the original way to do things or have been doing it 'this way' for quite some time. Since our industry is basically a subcontracting industry, the people you are introducing the change with are unlikely to be your company's employees. This limits your familiarity with them as you only have a short amount of time with them to find out their capabilities and how they actually work.

What causes people to be so negative? Since virtually everyone has to meet the pressures of time, specification and costs, there is little time to go for deep thinking on improvements.[18]

Often the four most common reasons, often gut reactions, for rejecting change are:

- Don't Understand.
- Can't Cope.
- What's in it for me?
- Not invented here.

There is more on resisting change and dealing with negative feelings on the CD.

Why people feel this way and how we sugar coat the solution is explored in the next chapter.

17 Sir John Egan expressed his disgruntlement on this aspect of our industry at the 10th anniversary of his Rethinking Construction Report. See Chapter 18 for more on Egan.

18 We are often much better at designing a sustainable building and building it traditionally than building in a sustainable way.

The final item on the implementation 'to do' list is to do something about potential problems. During any implementation, there are bound to be 'things' that can delay, distract, derail people from making progress. It can be people, as we have just seen, but can also be procedures, workplace layout, work schedule and general practices.

These are:[19]

- **People**: we know people can be awkward.
- **Methods**: how we do our work: procedures, practices. It may be a method that has been in use since the pyramids, but it is not necessary helping us to making sustainability happen. And do not forget – a lot of procedures are made up by people for their own benefit, not necessarily for you to do your job.
- **Measures:** if you have poor communications with your clients, then there is a good chance that you will be measuring progress and achievements on different sets of benchmarks. If you have different visions of sustainability, then problems can start.
- **Materials**: the raw materials, whether it is bricks, data or knowledge, may not be right to get the job done. For sustainability, it may be non-FSC compliant timber or no plans for reuse or designing out waste.
- **Tools**: our skills, the equipment and plant we are using may not be right for the job.
- **External event**: such as a credit crunch or even traffic jams can mess up many plans.

We know from experience that things will go wrong when we implement just about anything, however, we can assess what needs ot be done to prevent problems (see Figure 15.15) through the Failure Modes and Effects Analysis (FMEA) described earlier.

19 These are known as the common causes and are often used in a cause-effect (Fishbone) analysis.

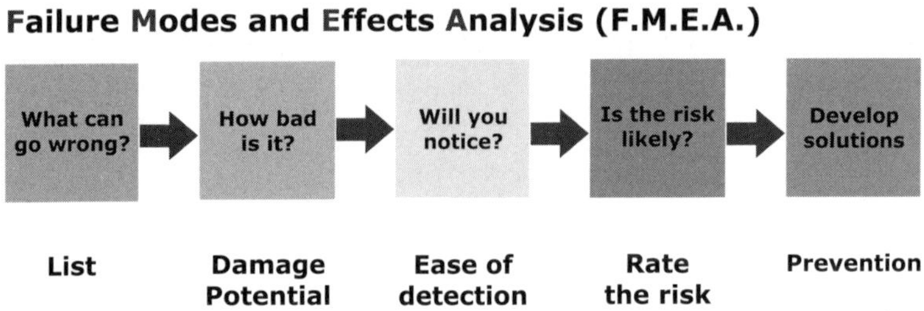

Figure 15.15 Problem prevention

As we found before, FMEA has four components:

- What can go wrong? This produces a list.
- What is the impact when it goes wrong? This allows prioritising the impacts.
- How likely is it to go wrong? This also allows prioritising.
- How easy is it to notice if it goes wrong? This provides an inspection routine.

It is not always easy to decide what goes on the FMEA list of possible problems. It may be possible to use the six Common Causes, listed earlier, to show what can go wrong. The hardest component to assess is the ease of detection. If it is not likely to be detected readily, you are also just as likely to miss it off the initial list of problems. If you include literally everything that can go wrong, you will have a lot of false alarms to distract you.

A possible methodology of problem prevention is suggested below.

1. Map the improved process backwards as in the 'pull' process and list the possible activities for each step (*'what needs to be done in order to deliver this?'*).

2. Categorise these enabling activities into the six Common Cause categories from the Fishbone analysis. The choice of the categories is not rigid and you may have your own preferences beyond (or instead of) these, but the six Common Causes cover most circumstances.

3. List what can go wrong in each of the cause categories.

4. Estimate the damage potential of each (a simple way is to rank them from 1–5), then comes an estimate of how hard it is to detect when the problem happens (again, rank 1–5), then estimate the probability that the problem will occur (rank 1–5). You can then add the score for each potential problem and rank them accordingly.[20] Then consider what solutions can be put in place to prevent these problems from coming up. It may mean a redesign of some aspects of the improvement.

5. Assign someone to monitor the potential problems. This is a fail-safe mechanism and can be effective for the hard to detect problems.

An alternative methodology is to say: this is too much bother. We'll skip it and firefight when it happens. As Sir John Harvey Jones, a previous chairman of ICI, was reputed to have said, 'Planning is an unnatural process; it is much more fun to just do something. And the nicest thing about not planning is that failure is a complete surprise rather than being preceded by a period of worry and depression.' There you have it!

Control

By analysing what can go wrong with the implementing the improvements, you have established a procedure for managing the activities under the control element of DMAIC. Both Lean operations and Six Sigma have very strong elements of monitoring progress and making sure there is no backsliding from any achievement. In 5S, there is the 'sustain' element and in Lean generally, continuous improvement is a mantra.[21] In Six Sigma, one element of DMAIC is simply called 'control'.

Controlling the gains sounds much less exciting than solving waste problems, putting together improvement plans or working out a sustainability policy. However, this is *the* crucial element unless you enjoy explaining to your boss why all the gains from your improvement project eroded to nothing.

20 You can, of course, weight these elements to emphasise their importance. The ease of detection factor may warrant its score to be multiplied instead of added to the others.

21 In quality management, it is the feedback loop or the 'CA' in the PDCA or Deming cycle described later in this chapter.

Control has several parameters and these are:

- managing and monitoring progress
- ensuring compliance
- setting targets, and
- continuous improvement.

Training in the form of tool box talks can be used to help these activities.

1. Control: Managing and Monitoring Progress

This is the detailed part of the control activities. It is about the daily grind of keeping up the effort, cajoling people and logging progress. It is labour intensive and frankly, a bit boring. It is intensive because regulations like the Site Waste Management Plans require regular monitoring of material flow and that means a lot of leg work. It is boring because progress is likely to be slow on a day-to-day basis. At the end of a month, you may start noticing gains but these are not readily visible when you march up and down the site every day. An easy way to observe changes over a period of time is a 'then and now' comparison. Photographs of the workplace before the improvement and then weekly photos of progress may raise morale as progress becomes visible.

2. Control: Ensuring Compliance

This is the 'policing' part of the control, but it is not about naming and shaming as that generally builds more resentment than it brings benefits. It is about combining cajoling with supporting. Generally, the aim is to help the teams on site to make sustainability happen. This means clearing away the hassles that make it difficult or in some cases, impossible, to carry out the sustainability improvement.

The FMEA analysis from the improve element of DMAIC provides a number of tasks and activities to ensure compliance. Potential problems that may cause non-compliance will need to be dealt with. There are also the possibility that the FMEA analysis may have missed some other potential problems especially if these are influenced by external events. Then you will need to be active in exploring the way out of those problems. This brings us to targets and flexibility.

3. Control: Setting Targets

If you are asked to set sustainability targets in early 2009, it is very likely that they will be influenced by events that hit our industry. This means that, although sustainability is important, corporate survival is probably more important in the short term especially if you are interested in getting your salary paid. This can mean revisiting the targets set in a different business environment.

Setting meaningful targets is never an easy task. Even though target setting is not considered to be as hard as rocket science, it does require some care and planning. Targets need to be motivating, somewhat stretching, but ultimately doable and relevant to the corporate needs. A simple way of making sure targets make sense to the project (see Figure 15.16) is to deploy the SMART approach to target setting.[22]

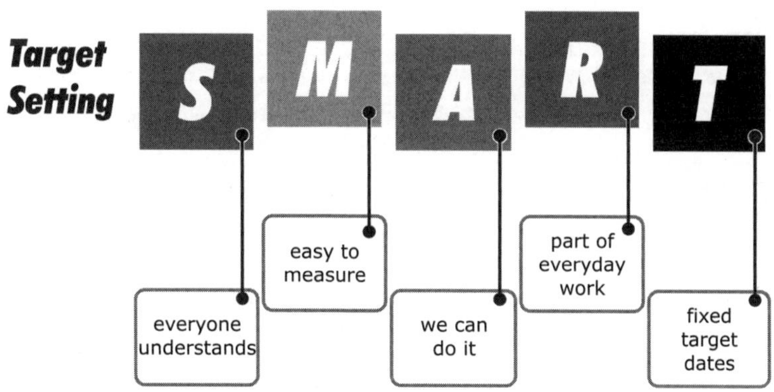

Figure 15.16 SMART target setting

22 I introduced this earlier as a set of filters for observation in Chapter 7.

SMART targets are:

Specific – the targets are clear and not vague so everyone understands.

Measurable – you can easily measure these outcomes as quantitative data.

Achievable – the targets are not impossible – we can do it.

Realistic – reaching these targets do not require specialist training and can be part of normal everyday work.

Timed – the performance is measured within a fixed length of time: target dates are set.

'Specific' – a specific target means that it is quite clear what you are measuring.

Table 15.7 Specific targets

Specific Targets	Not Specific Targets
Get the pipes to the site by 3:30 PM	Get the pipes to the site sometime today
Everyone to pass CSCS health and safety test	Everyone to know something about health and safety
'I'll have egg, bacon, chips and beans'	'Gimme a fry-up'

'Measurable' – results and outcomes that can be measured.

Table 15.8 Measurable targets

Measurable Targets	Not Measurable Targets
The project has achieved ISO 14001 certification	We take environmental management seriously at this project
We reduced our plasterboard waste by 22% (by weight) compared to the last project	We threw away fewer sheets of plasterboard this time
The entire team achieves a minimum NVQ level 2 standard	Our training is about the same as an NVQ level 2

'Achievable' – setting sensible targets that the teams can actually achieve.

Table 15.9 Achievable targets

Achievable Targets	Not (Really) Achievable Targets
Our target is to reduce rework errors by 35% compared to the last project	We will have zero defects
We established weekly review of environmental incidents	We will not pollute on this project
Our project plans have some 20 days of contingency	Our project will run to cost, be on time and 100% to specification

'Realistic' – targets need to have real for the people involved.

Table 15.10 Realistic targets

Realistic Targets	Unrealistic Targets
It will take about 2 weeks to deliver the sustainability message through tool box talks to every one of the 500 people on site	We'll put sustainability posters up in the canteen and dry room, everyone get the message within a day
We expect to return your calls within 1 working day	We will return your calls within 1 hour
We will pay your invoices in 30 days	We will pay invoices on receipt

'Timed' – targets lose their value if there is no time limit on when they need to be achieved.

Table 15.11 Timed targets

Timed	Not really Timed
The second floor fitted out by 24 May	The second floor fitted out sometime in late May
The project will start on Tuesday and finish by next Monday	The project will start sometime this week and finish sometime next week
We will need 4 hours of crane time this Thursday, 10–2	We'll need about a half-day of crane time sometime later this week

SMART targets provide a project with more precision and clarifications, and the SMART approach to target setting is fine in itself, but it is also tends to be set in a self-contained world where nothing external is taken into account. It would be fine to set SMART targets if the world remains at January 2007 or whenever/wherever the credit crunch is not. But with national governments everywhere slashing interest rates and the global economy in a depression, would these targets still be right for the job?

In a recession or when the business climate changes, we need to make sure the actual target values are changed to reflect the changing needs of the company, but they remain SMART targets, even if everything aspect has changed. Of course, if you get the new targets wrong then you will look bad by having to change them over and over.[23]

Although SMART suggests we go for targets that can actually be achieved, this is an area of some dispute. Some managers believe sustainability ought to have tough targets to make people try harder, others take the opposite view: easy targets to avoid disillusionment. Common sayings, such as 'aim for the sky' and 'low-hanging fruit', offer conflicting visions of effective target setting. My view on this is quite simple: a low-hanging fruit in hand is something you can eat now.[24] Easier targets mean more people can get involved and achieve them. This then allows people to overcome their fear of something new and different. The aim of sustainability in construction is to engage everyone in doing sustainability, therefore, setting targets that only graduates with rocket science degrees can achieve is self-defeating. Nevertheless, staring with easy targets does not mean staying with easy targets!

Tough targets are best for an experienced, long standing and close-knit team. These teams know their capabilities, maybe even their Belbin team roles and are therefore familiar with complementing and supporting each other. For scratch teams put together from different contractors for every new construction project, it may be prudent to start with easier targets to bring everyone onto the same page (a backwards or sideways step for some, no doubt). A main contractor may wish to raise the sustainability target levels from project to project but the success depends on training, support and consistency in the composition of the subcontractor teams. The framework approach is a situation

23 Embarrassing examples of extending dates are all around us. The delays during the building of Wembley Stadium is a good example.

24 Oops! This is getting close to the short-term outlook I described earlier about getting management commitment to sustainability. However, if you need everyone to get on board, then you need to set the platform at a reasonable height.

where all parties can gain familiarity with each other over a period of time and therefore more able to deal with stretch sustainability targets.

Target setting, therefore, ought to be a partnership activity, with the team, the supervisor or manager and the specialist discussing what is achievable right now; what is achievable with some support; and what is achievable with extra experience (see Figure 15.17). After everyone is clear on the support available and the level of experience of the team, then it is time to agree a target that everyone can participate actively.

Sustainability is not a winner take all race, it is a mass participation activity that can declare victory only when everyone is committed. And to get there, we need to improve continuously.

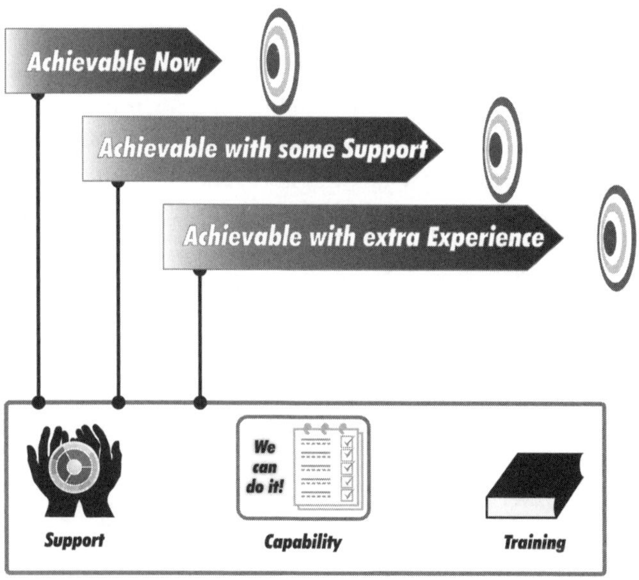

Figure 15.17 Target settings

4. Control: Continuous Improvement

Although Lean and Six Sigma both shout the importance of continuous improvement, there is a danger of going to the extremes, of missing the big picture and seeking perfection to the ninth decimal place. Using Six Sigma (3.4

defects in 1 million opportunities) as an example itself, we can clearly see that continuous improvement is a relative thing.

- If there are 35 blocks somewhat misaligned in a 10 million block array (think the Great Wall of China), how much will getting those last 35 blocks lined up cost? Is it worth the expense?

- 70+ industrial deaths in 2 million worker-years (think the UK construction sector in 2007), that is an industry which is about 99.9955 per cent death-free. Is this a target for continuous improvement?

Continuous improvement works when we feel we can do something about it, and also that it makes sense to do something. The cost and effort in aligning the last 35 blocks may not be worthwhile but reducing our industry death toll is definitely worthwhile. Sustainability itself is a process of initial effort followed by continuous improvement of what has gone before. Like the initial effort, it is dependent on a number of factors, such as budget, commitment, skills, knowledge and time. The actual process of continuous improvement is more or less the same as the initial improvement. You ask, you observe and you review, then agree new targets and then get on with it, and so on and so on. This brings us to the familiar PDCA improvement cycle (also known as the Deming cycle, after the quality guru) – plan, do, check and act.

Table 15.12 Plan, do, check and act

Plan	What are we going to do? (benefits, risks, training, resources, schedule)
Do	Implementing the change (motivate, cajole, train, support, reward)
Check	How did we do? (measure actual against expected gains)
Act	Let us improve on it (analyse weaknesses and develop solutions)

The reason why a lot of continuous improvement programmes falter is because they are not continuous. Companies are attracted to the fashion of the newest 'black' and then quickly abandon it for the next corporate fashion item. However, sustainability is backed by legislation and that never goes out of fashion (it will only get more stringent and onerous!). Budget, commitment and time are all resources in short supply when the economic situation takes a turn for the worse. So it comes down to using the available resources to keep the initiative flexible and on-going. It is time to deploy the talent, skills and knowledge of your entire employee population to make things happen. That, and a dose of respect, recognition and rewards will get you a long(ish) way.

5. Control: Training Through Tool Box Talks

I described how tool box talks can be used to help teams to come up with solutions. It can also be used to help control the gains from the improvements. There are several ways you can use a problem-solving tool box talk session to make sure the gains are not lost.

- Describe the eroding gains as a problem and ask the team to solve it.
- Treat the possibility of eroding gains as a potential problem and invite the team to come up with a preventative solution.

An additional benefit is that this shows the overall commitment towards holding on to the sustainability gains.

16

Respect, Recognition and Rewards: The 3Rs That Sugar Coat the Implementation

In a difficult business climate, it is often hard to keep both enthusiasm and momentum up when budgets get cut and colleagues get laid off. You need to walk a fine line between rallying your colleagues as well as not falling foul of the consensus where anything that does not generate income is viewed with some suspicion. Besides, you still need to implement the improvement ideas you and your colleagues worked so hard to set up. Since the implementation is probably your day job as well, let us see how the implementation can go forward (see Figure 16.1).

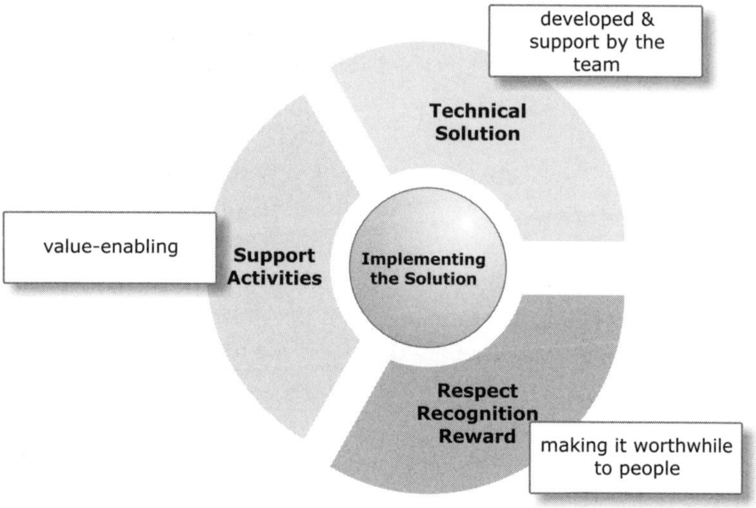

Figure 16.1 Sugar coating the implementation

There are several components to make the implementation successful. These are:

- a sound technical solution;
- support activities for the implementation; and
- respect, recognition and rewards.

Technical Solutions

I have covered how to get to a technical solution in the previous chapter and I mentioned that a good technical solution needs a sugar coating, and this chapter is about getting people to accept the change and then actually change behaviour as the sustainability improvements are implemented.

Support Activities

In Chapter 8, I identified a major component of our work, the 'value-enabling activities', that allow us to makes things happen. These can include training, on-the-job guidance, troubleshooting and encouragement. It is essentially a case of breaking logjams, deflecting bureaucracy and getting rid of hassles for the team so that they can focus on the carrying out improvements. The control element from DMAIC will mean some type of 'checking system' is in place so if the 'most likely cause' agreed by the team turned out *not* to be the most likely solution, you can catch it quickly!

Another way to work out what support activities are needed during the implementation is by using tools, such as the FMEA (see Chapter 15). Potential problems can be listed and the necessary support activities to make things run smoother are part of the solution. Nevertheless, probably the most effective support activity is through motivating the people doing the work to make sustainability happen. This is where the 3Rs come in.

Respect, Recognition and Reward

Reward is a touchy subject because a lot of teams will automatically feel that if they have to do anything different, they are entitled to some form of reward (leaving respect and recognition for the moment, as only reward pays the bills). It is also touchy because many companies will automatically feel that the workforce should deliver higher productivity for the same pay.

Both these approaches exist in our industry but for sustainability to happen, we need to move away from these extremes. Yes, we believe that respect, recognition and rewards are important, but we also accept that sometimes, there just is not the money available for the rewards.

The outcomes of a sustainability improvement may be about changing business practices but underneath it all, it is often about changing people's behaviour. Whatever the slogan says, it is really:

We want you to do it like this, or

the new approach is about doing it that way.

This looks like opening a whole new can of worms. Since sustainability belongs to everyone, can we really ask people to change their behaviour? Actually, yes, we are using the company's Core Intent statement to set the direction, provide tools and offer people the opportunity to make sustainability happen.

If, for example, the Core Intent is 'paying more attention to safety at work', the company then provides training on tools, such as Value Stream Mapping (to highlight potential and actual safety black spots), as well as the cause-effect and root cause analyses (to examine and deal with the safety black spots) to help the teams improve safety at work. The workforce uses the tools to deal with the safety issues and subsequently they own the solutions. Nevertheless, when they implement the solutions (see Figure 16.2), they will still need to change their behaviour to make the improvements happen.[1]

Therefore, however simple the improvement, people still like to be rewarded for their behaviour change. However, 'reward' does not have to be money, reward, in the general sense, can include many others besides cash or

1 This is partly human nature, as many of us often dissociate the planning and thinking (problem solving) from the actual doing (implementing).

gift vouchers. Reward can also include respect and recognition. In view of how this industry sees money, we will list these options starting with the free stuff (see Figure 16.3).

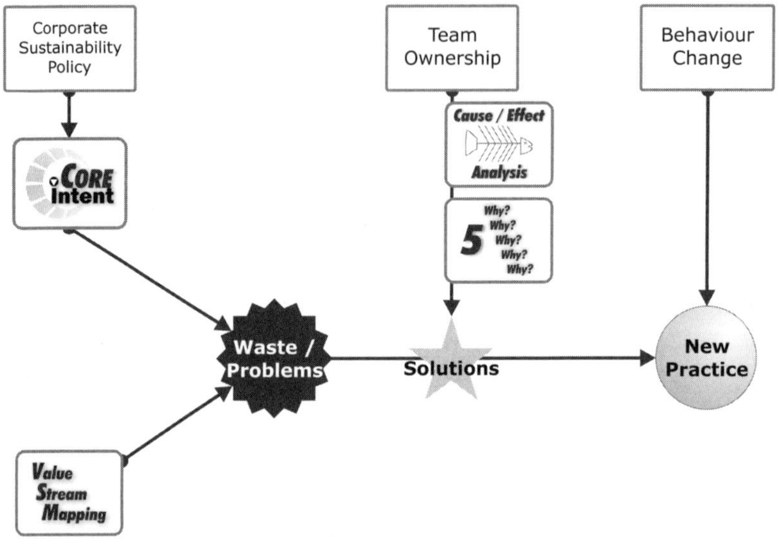

Figure 16.2 From new policy to behaviour change

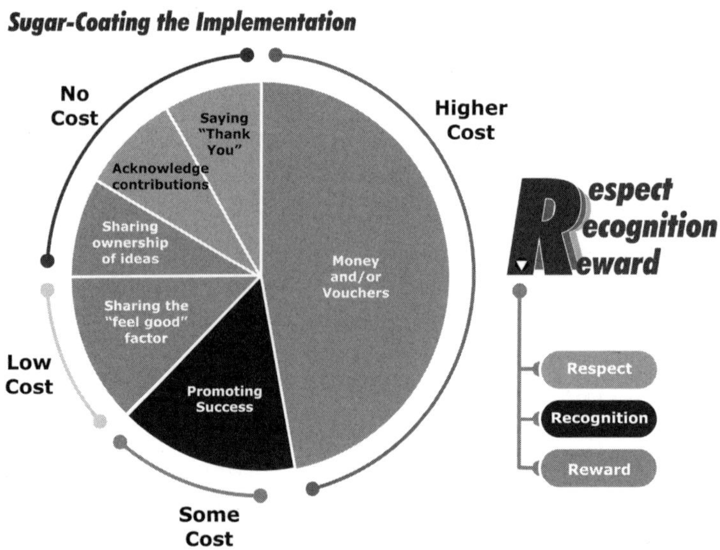

Figure 16.3 Respect, recognition and reward

NO COST – SAYING 'THANK YOU'

This is very easy to say and although it may be only a small gesture, it will make somebody's day if it is done sincerely. It is one of the most powerful tools to make sustainability happen. 'Thank you' is about appreciation and it is a sign of respect. It also makes the people receiving it feel better by increasing their self-worth.

NO COST – ACKNOWLEDGEMENT

This is a bit more than thank you, it is a statement that says, 'by doing this specifically, you have made a difference'. The specific nature of an acknowledgement means that the individual can link the appreciation shown to specific actions. The cynical ones among us will start thinking of behaviour tampering,[2] why not? We want to respect, recognise and reward positive behaviour.

NO TO LOW COST – OWNERSHIP

When management imposes a solution, the ownership is not shared and people tend to care less for the things they have no say in. But when the team comes up with a solution themselves, they feel some ownership towards the solution. (Of course, not every activity can be managed that way, health and safety rules, for example, are not things that can allow such flexibility.) Encouraging teams to come up with solutions to waste problems and then implementing the solutions do provide a sense of ownership. Even better, the team will defend the decision until you invite them to improve on it the next time. Sustainability happens a lot faster when everybody has a piece they can call their own.

When a team innovates and delivers a new way of doing things, people often feel that they are not just interchangeable units of labour, but individuals that have some control over how they work. It is a feeling of pride, of being recognised to have the ability to contribute towards the way work is done. This is highly motivating and cements the feeling of being engaged and involved in the sustainability improvement.

2 Remember Pavlov's dogs? Ivan Pavlov discovered that reflexes can be triggered by conditioning. Pavlov trained dogs to associate food with the sound of a ringing bell to the extent that when the dogs heard a bell ring, they salivated as though food was also present. Find out more about Pavlov at the Nobel Prize website – he won his in 1904.

What are the costs of this? Other than spending some effort and time training teams so that they can come up with innovations and then coaching and facilitating them, there is not a lot of other costs. (A often forgotten point is that when a team come up with an innovation, it is usually low costs. This is because most people in the team will have no budget authority, therefore big ticket items are less likely to be considered.)

LOW COST – FEEL GOOD

The media is a driver for sustainability and we often spend too much time reading and talking about sustainability instead of 'doing' sustainability. A very motivating reward is being told specifically that your activities have 'done good'; the feel good factor in knowing that you made a difference is immense and very difficult to quantify.

It can take a bit of work to figure out how exactly it is that we have 'done good', and then to link our activities to something which we can relate to as people and not just as construction professionals. This is an activity that has a cost as it requires time and effort on somebody's part. Once it is established that the team made a difference, that is it. Instead of talking about sustainability or reading about it in the press, the team is actually doing it. They are making sustainability happen and that feels good! This is especially so for waste reduction – when you reduce waste, then you are doing good, doing sustainability *and* saving money!

SOME COST – RECOGNITION

Simply put, this is a public proclamation of someone has 'done good'. You put the team's achievement in a case study or on a poster at work and making sure that their photos are there too. Sending copies to the local press and media is also a sign of recognition. It is saying that the company is proud of its people, so proud that it wants the world to know about it. For subcontractors, it feels good when their bosses gets a note of thanks from the main contractor highlighting the achievement of the team. For the individuals, it feels good every time they walk past the notice board. If you have done something to make sustainability happen a bit quicker, this really motivates you to do more. For teams which have not got their act together yet, it is motivating because they know they are as good as you and their chance will come soon.

The costs here are the promotional costs, but since it is for doing good and makes everyone (from neighbour to client) look good/feel good, few people will mind the cost.

HIGHER COST – MONEY

Money is a blunt instrument in motivating people and changing behaviour, but it can be effective if the rewards are shared openly and fairly. The level of subcontracting in construction can make this a very difficult activity to carry out. In many other industries, any cost savings are shared between the various parties involved. In construction, the movement of contractors may not coincide with the timetable of the calculated cost savings and therefore the smaller subcontractors may lose out as they may have moved on to different contracts. One way to overcome this is by having a formal profit-sharing agreement, such as Gain-Share in the contract. This is discussed in more detail later.

Since people can always think of ways to spend money, money is always welcomed. The problems start when people expect extra money as a reward even when they are doing nothing special. Expectations also rise (and probably faster than the rate of inflation) so people end up wanting more money for doing the same thing.

ENGAGING PEOPLE

If there is absolutely no money, then what have you got left. A silver tongue perhaps? Motivating people solely by your sincerity and charisma is not a long-term solution because the impacts disappear when you are not there or when the team feel like they face an endless stream of barriers and difficulties.[3] However, developing a sustainability partnership is something different. This is about encouraging the teams to come up with ideas. The reasoning behind this is simple: since everyone has an idea about sustainability, then by listening to them, we are showing respect and recognition. By relating what they do in the context of our sustainability objectives, we are telling people they can make it happen. By encouraging people to come up with ways to improve the way they work and reduce waste; we are giving our teams a degree of local control over how they work. If we do not make use of this extra (or 'free') resource, then

3 If no one is offering any budgets for rewards, then it is a sign that there are likely to be structural barriers as well as cultural barriers. The no cash reward culture does hamper innovations, which is often the reason why companies that do not reward (including the public sector) are rarely exemplars of innovation.

we are creating waste in talent and ideas. And this is one of the most significant wastes in any form of work.[4] Everyone is unique and therefore the ideas are also unique and valuable. Many organisations have no effective mechanism for dealing with the ideas and innovations except through suggestion schemes. Often these are discredited with an opaque reward system and do not come with either respect or recognition. The approach I developed for the NoWaste programme gets around these issues through the following four consecutive stages:

1. An invitation to participate – at the induction, everyone is advised that they will be expected to take part in waste reduction on site.

2. Demonstrate an understanding of the constraints – not everyone knows about sustainability or has knowledge on waste reduction.

3. Provide support – training in the form of special Tool Box Talks (see CD) as well as the support of the on-site sustainability officer to encourage discussions on ideas for improvement.

4. Respect, recognition and reward for the ideas generated.

Not valuing the contribution of others is an arrogance that some people display. This is probably because they think they have all the answers, whereas in reality they are letting innovation and competitive advantage slip through their fingers. People doing the work often have ideas on how the work can be improved, the only reason they do not tell you is because there is often no relevant mechanism for them to discuss their ideas and innovations. But think of the motivating feeling you get when your idea for improvement has been implemented. Suddenly, you have changed the way work is carried out and you realise that you do have some control over the way you work. If your idea was not accepted, at least someone came and explained why to you, showed you respect and acknowledging your contribution.

One of the ways to help motivate people, other than engaging them in generating ideas, is a money-based reward system that is open, transparent and broader in scope than just money in your back pocket. It is about making real achievements and then using the reward not just to fill back pockets, but also to do good. This reward system is called Gain-Share.

4 See the Eight Hidden Wastes in Chapter 8, Figure 8.6.

Gain-Share

Gain-Sharing[5] is a strategic rewarding system that allows every partner to win through working together. It is not the same as a bonus reward scheme because Gain-Share is based on the value-added or productivity gains rather than arbitrary performance targets.

Gain-Share has an interesting history, it was invented in the 1930s by Joe Scanlon[6] in the USA. The original versions of Gain-Share were called Scanlon Plans and they comprise a mix of cost-accounting, partnership and continuous improvement. The US industry then was dominated by the thinking of Frederick S. Taylor's scientific management or 'time and motion'.[7] Scanlon pioneered the engagement of the workforce to deliver continuous improvement solutions.

The process works where a company and its stakeholders (one group such as employees or multiple groups, such as business partners, employees and community) agree a sharing of the 'gains' from business improvement. The gains elements can be a single element, such as increases in sales or productivity, or a multiple of financial and non-financial gains. In addition, the gains can be short, medium or long-termed elements.

In simple terms the Sustainability Gain-Share I am proposing can be described as picking up currently wasted efforts through working smarter. The important thing to remember about Sustainability Gain-Share is that the gains are mainly recovered from waste reduction (see Figure 16.4). Therefore when the share-out happens, companies should not need to worry about 'giving money away' since the money was originally lost as waste anyway.

---

5 Egan recommended Gain-Sharing as a way forward in his 10th anniversary review of Rethinking Construction. *Building Magazine* – see Chapter 17.

6 Joe Scanlon was variously a cost (or management) accountant, a steel worker, a union organiser, a university lecturer and a prize fighter! Apparently he regularly showed up at the accountant's office black and blue with bruises from the previous night's fight, bringing a new meaning to the phrase 'multi-skilled'!

7 Frederick Taylor's time and motion studies were actually about improving productivity by aligning the right person to the right job. But his methods and approaches were contrary to modern working practices. Find out more by downloading Taylor's book: *The Principles of Scientific Management*, free from the 'Project Gutenberg' website.

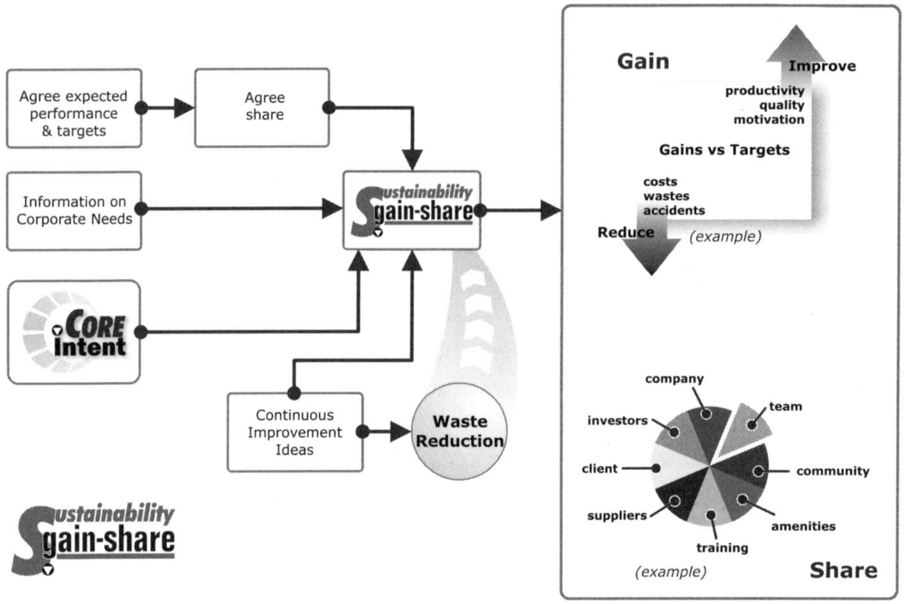

Figure 16.4 Sustainability Gain-Share cycle

To kick off a Gain-Share programme, companies need to provide its teams with information on itself – why are we doing this? What is it we wanted to see happen? How this is important? Information is the key to get people involved and engaged. If our colleagues do not understand why they need to do sustainability (let alone how to do it), it is unlikely you will get much enthusiasm for your initiative. Some companies will find this hard, since corporate figures are often treated like state secrets. Luckily for those companies, sustainability is changing the rules. Some information needs to be shared so that your teams can put their heart in the sustainability improvements and transformation.

The Gains: Measures For Added Value

One of the most difficult parts in determining a set of 'gain' elements is to consider which activities add value. This is can be examined through the concept of work being divided into three components: value-adding work; value-enabling work; and waste, discussed in Chapter 8.

In the construction industry, there is another way to spot where gains can be made – the Site Waste Management Plan Report lists the predicted waste levels. If this is not a starting point for reducing waste, then what is?[8]

Value-adding work is about functionalities for the client and this is quite easy to determine. Value-enabling work is less easy to determine because this type of activity is not linked to customer utilisation. The Lovelace List provides some idea for the value-enabling activities provided it is pragmatically applied. Unbalanced value-enabling activities can actually distort Gain-Share measurements if they are not designed properly. For example, a gain target of 'increase well-being by X per cent' can be meaningless if it cannot be associated with 'costs of accidents' or 'reduced errors' because without a practical measure, the investment in effort may not be easily assessed and you are not likely to win support for the Gain-Share programme. This is when each element in a Gain-Share programme needs to be assessed carefully. Simple gains, such as profit, is easier to assess although attributing it to an individual suggesting good ideas may take some work to figure out.

In some instances, it may be easier to measure the amount of waste reduced rather than look for exact gains. For example, a ten-step procedure may be reduced (safely) to seven steps with a corresponding reduction in time and resources. The time saved can be calculated in financial terms and be equated to a cost. Then the gains can be readily assessed. Generally, a management or cost accountant can work out the cost of an individual's time (not the wages, but the total cost of employment) and similarly work out the cost benefits of a new set of activities compared with the old set. This allows the 'monetisation' of effort and gains, and makes it easy to calculate what the total gains are.

The type of element that can be accompanied into a Gain-Share programme can include (the following is only a sample list):

Table 16.1 Elements of a Gain-Share programme

Accident rate	Cash in hand	Customer satisfaction	Employee satisfaction	Legislation compliance
Margin	Order book	Productivity	Profitability	Sales revenue
Schedule adherence	Supply chain efficiency	Stock market value	Waste levels	Quality

8 Reality check time: what really happens is a bit different – perhaps the onerous nature of the SWMP form filling made it more a tick-box activity instead of an inspiration for innovation.

Every company will have additional elements they can add to the mix of gain elements. For example, environmental and social sustainability elements (see the Lovelace List in Chapter 13) can be added to mix.

Managing a Sustainability Gain-Share Programme

There is a crucial issue in setting up a Gain-Share programme and this is the building of respect and trust between the employer and the employees or between the company and its business partners. The reason for this is because there is plenty of room for unethical dealings in managing a Gain-Share programme. A likely negative action is a withholding of the amount being shared. This is often a 'knee-jerk' reaction when a company realises that it will need to share a substantial amount of the gains: *'we can't give that much money away'*. Even though this money was previously lost through poor operations or waste. Another is to withdraw the programme after one cycle of improvement, mainly because of the amount of gains being shared seemed high. The result of both these is the same, an erosion of respect and trust leading to the drying up of new ideas and all the gains are subsequently lost.

The Gain-Share essentials are:

Engage people by consulting early and often – extensive consultation with clients, suppliers, employees and relevant stakeholders means that everyone's interest will be taken into account and everyone will be able to understand and buy into the Gain-Share programme.

Variety – have a right mix of gain elements. While it is more difficult to manage, a good mix of financial and non-financial measures as well as short and medium-term elements provides flexibility and opportunity for all partners to understand what is value for the particular company.

Fairness in administering the Gain-Share programme means that there will always be some reward for delivering improvements in added value.

Good **communications** – regular meetings between the Gain-Share managers and the programme's stakeholders and making sure that the minutes of the meetings are always easily available. Ensure that regular feedback is given on the progress of the Gain-Share programme.

Gain the **commitment** of senior managers for the Gain-Share programme, whether at the site/project or at the corporate level.

Establish a **timed** Gain-Share programme cycle with a clear and relevant start and finish times. Making sure that the cycle period meets the need of the programme stakeholders.

Openness is the foundation to gaining respect and trust in a Gain-Share programme.

Setting Up a Sustainability Gain-Sharing Programme

The elements in the 'gains' were explored earlier together with the 'essentials' list above, but a Gain-Share programme is more than creating a list of areas to check for improvements.

Information sharing in a Gain-Share programme allows everyone to act accordingly, based on facts. Now this may sound like an advert for Six Sigma, where decisions are based on facts, but the openness requirement and the numbers of partners involved in Gain-Sharing often mean that unambiguous information is essential to avoid misunderstandings.

Often it will be beneficial to share relevant information on corporate performance with the workforce. Gain-Share relies on people having the necessary information so that they know which areas need improvement. Therefore information on how a company is doing allows people to understand the reasons why the gains are important.

Nevertheless, relevant information needs to be in context with the people doing the work so that they can use it to make Gain-Share a success. They need to know:

- Why there is a need to add value?
- What areas should they focus on?
- What is in it for them to add value?
- What is the programme timetable?

Linking Gain-Share with corporate performance information is the Core Intent statement. This allows the company to distil its message into a very simple and memorable form. It is both empowering but also acting as a perimeter because the Core Intent statement performs a checking function on the 'do/not do' decision, thus empowerment for the teams doing the job but limiting it within the corporate objectives. At the same time, the empowerment is constrained by the Core Intent statement – you do not stray beyond what is in the statement.

Although the need to add value may appear obvious to senior managers, the company still needs to define its needs clearly. (The House of Sustainability tool in the next chapter explores how needs can be defined.) This is to ensure that the Gain-Share targets are relevant to corporate needs. However, it also allows the company to coordinate the Gain-Share programme to a different or new corporate direction.

Sustainability Gain-Share – Running the Programme

Managing a Sustainability Gain-Share programme is like running a continuous improvement programme, a suggestion scheme and a bonus programme all rolled into one. The main difference Gain-Share has with all these schemes is that the level of gains and share split are openly announced and everyone knows the targets.

1. With information comes awareness: the teams and individuals realise that any improvement will mean something definite in their pockets and perhaps something towards doing good as well.

2. The Tool Box Talks can be used not just to spot problems and develop solutions but also a Gain-Share event. This forms the improvement aspect of continuous improvement. All the parameters of managing innovations are applied with respect, recognition and rewards.

3. The solutions are implemented and the Gain-Share administrators log the progress against targets.

4. The gains are shared in as agreed in the original Gain-Share agreement.

Managing the Tensions

An agreement on how the gains are shared is important to make sure there are no lingering disagreements. What often spoils a Gain-Share programme is greed. One or more parties may resent other parties getting a share of the gains. A clear and open declaration of the share proportions coupled with a commitment to review the shares for the next Gain-Share cycle will go some ways to reduce the tensions, but the programme administrators will need to be aware of any undercurrents.

The concerns about greed is linked to a second problem with Gain-Sharing – the need for an unbiased administration of the programme. This is also critical because it helps to manage accusations of favouritism for one group over another. One way to avoid this is to appoint an administrator with no interest in the gains or the shares. An outside management or cost accountant[9] on a fixed fee may be better suited to administering the gain-share programme than in-house staff with vested personal interests.

Sharing the Gains

At its simplest, Gain-Share can be a rewards sharing scheme based on improved business performance past a certain agreed threshold (see Figure 16.5).

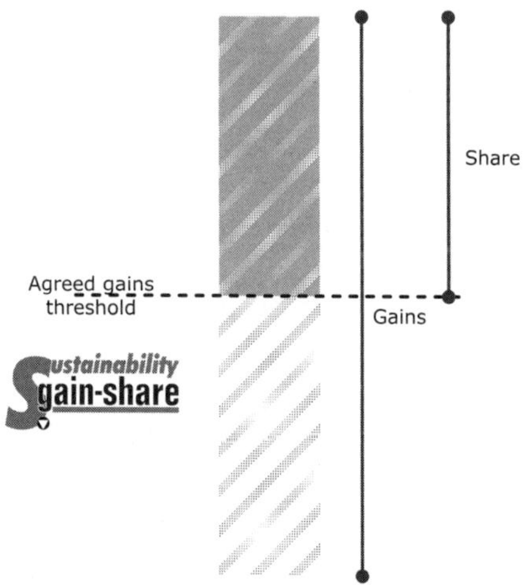

Figure 16.5 Gain-Share thresholds

9 CIMA is professional body in this area. Find out more from the Chartered Institute of Management Accountants' website at (www.cimaglobal.com).

More complex methods of sharing can occur between a contractor and a supplier. Nevertheless, based on an agreed set of performance targets, the gains can be shared variously. The following examples show different combinations of sharing. The detailing is an important aspect of Gain-Share because it demonstrate openness.

Example, in-company only option:

- 25 per cent to investors/shareholders;
- 25 per cent to company for retained profits;
- 25 per cent to corporate investments (e.g. plant, amenities, training, apprentices); and
- 25 per cent to employees of which 10 per cent goes to the team with the improved productivity, 10 per cent to the department which the innovating team belongs to and 5 per cent to all employees.

Example, business partners option:

- 25 per cent to suppliers;
- 25 per cent to customers;
- 50 per cent retained in company and shared as the in-company only option.

Example, social responsibility option:

- 50 per cent to company/business partners;
- 25 per cent to employees;
- 25 per cent to social and community projects.

Essentially the way the share is divided depends on a partnership between the company and its stakeholders. A more complex sharing option will require more extensive monitoring and managing.

Gain-Share works as a sugar coating to behaviour change because it is seen as essentially a fair system where efforts and results are linked to a share of the rewards.

REALITY CHECK: WHAT ABOUT SUBCONTRACTORS?

Since a lot of contracts are fixed priced, then surely any cost savings through improvements will go directly to the subcontractors and not the client or the main contractor. So how can a construction project manage a Gain-Share scheme and claw back the gains for everyone to share? This comes back to a matter of real partnership working – which is something that is rather weak in the UK construction scene. In theory, frameworks can help to a certain extend in sharing the goodies in the short to medium term. However, often we hear of the main contractor demanding cost reductions or subcontractors underbidding to win the contract. Clearly Gain-Share is not going to work when working relationships are not based on partnership and trust but more on trying to outwit each other via the contract terms. But when companies are prepared to work in partnership, an open and clear Gain-Share system will benefit everyone including subcontractors, main contractor and the client.

Behaviour Change

We have looked at how behaviour change can be motivated by incentive or reward, but how does behaviour change take place? Peter Willmott,[10] a consultant, developed a five-step behaviour change process (see Figure 16.6). In the description below, gaining 'information' or 'knowledge' are used as examples for stimulating change, but the behaviour change process can be equally valid for learning a new tool, technique or a Method Statement.

10 Peter Willmott, of Willmott Solutions, developed this as part of his Total Productive Maintenance programmes with Lynn Williams of Lunces Training.

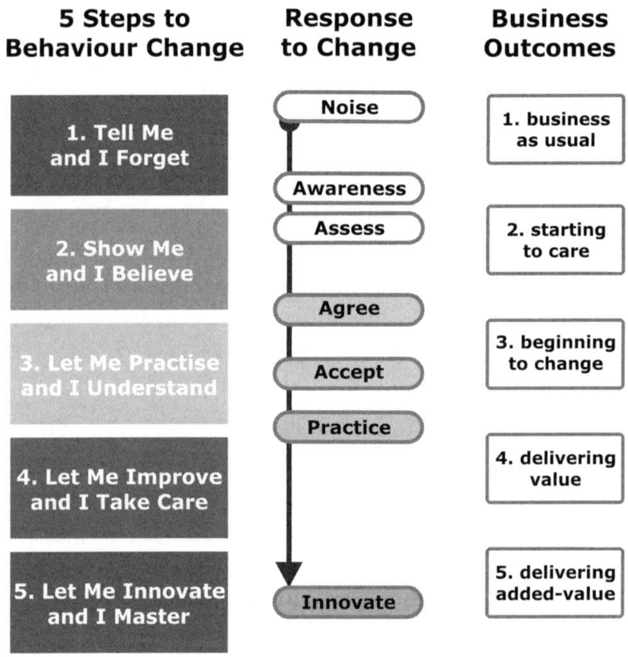

Figure 16.6 A five-step behaviour change process
Source: P. Willmott

STAGE 1: 'TELL ME AND I FORGET'

The first stage of the process is 'business as usual' or no change at all. It is best explained as the 'in one ear and out the other', a syndrome many in construction know well through the often uninspiring induction process. The information, however useful, is dull, boring and poorly presented.

STAGE 2: 'SHOW ME AND I BELIEVE'

However, if some care is taken with the message – put it into the right context, showing the audience respect and inviting their opinions. Then people will start to care because there is interesting knowledge that can be used.

STAGE 3: 'LET ME PRACTISE AND I UNDERSTAND'

This is not 'being told what to do', if everyone is encouraged and allowed to apply the knowledge in work situations, then the whole thing will start to make sense. With understanding comes reflections on how this knowledge can be deployed in other situations. This can lead to an even deeper understanding of how to get the most of the deployment and changing the way work is carried out to enable greater effectiveness.

This works very well if the initial deployment is successful. Since everyone likes successes, if the deployment worked the first time, then there will be some enthusiasm for additional attempts. However, much of this is lost and the situation reverts to Stage 1 if the deployment does not work, so carry out a trial or two before rolling it out to everyone.

STAGE 4: 'LET ME IMPROVE AND I TAKE CARE'

This is when people gain confidence in using the knowledge and starts to think about how this knowledge can be improved. At this stage, people are becoming empowered and start to deliver value through using the information. Behaviour change has taken place and the individuals are in a position to help others to use this knowledge.

STAGE 5: 'LET ME INNOVATE AND I MASTER'

Not everyone reaches this stage but this is when familiarity with the knowledge has resulted in a deep understanding. The user is ready to develop completely new knowledge and this is when added value is delivered. The company gains a 'guru' and it also extends its competitive lead.

Behaviour change is hard work but possible, and it needs commitment from both the organisation and the individual. However, the fruits of behaviour change are many: reduced waste; improved morale; and reduced costs. Does this sound familiar? Yes, we are 'doing' sustainability when we encourage, enable and empower behaviour change that is consistent with the corporate sustainability Core Intent.

Behaviour change is hard because it involves change. As we saw from the simple exercise in Chapter 15 that even crossing our arms in the opposite way is hard. This difficulty is often called the Switching Cost, it is the tangible and intangible costs of changing from one activity to another.

The Switching Cost

Sometimes people avoid change because it means spending a lot of money. For example, setting up a sustainability programme means paying for a system to be set up, equipment to be bought and people to be hired or trained, as well as paying someone to audit the entire process. But it can also be an emotional cost and this is the much harder cost to measure, quantify and overcome.

Lynn Williams of Lunces Training, suggested that the Switching Cost decision is based on the assessment of available information and trade-offs (see Figure 16.7). The reason for a 'refusal to switch' response is that with insufficient information, people often tend to think the worse of things. This is made worse with unfamiliar subjects, such as doing sustainability, where not many previous benchmarks are readily available to the individual. Recent research also showed that requests, which are vague, tended to result in procrastination,[11] and 'do sustainability' is often as vague as they come.

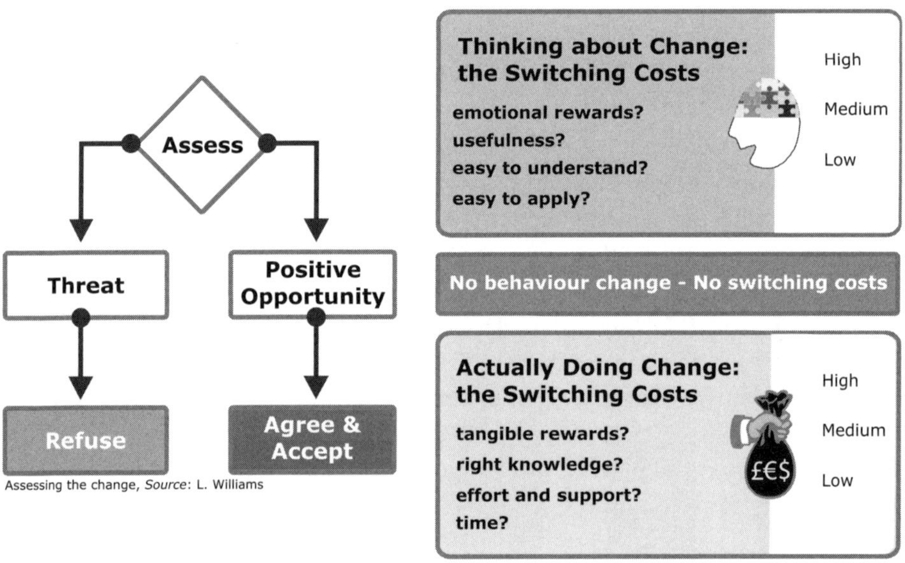

Figure 16.7 Engagement in switching costs

11 'Motivating Minds', *The Economist*, 24 January 2009.

The Switching Costs has two components (see Figure 16.7), just like behaviour change. There is the *thinking* part where the brain thinks about a new way of doing things, and there is the *doing* part where the individual gets on with it in the new way. Both have their respective Switching Costs and although they are related, these are still different. I explained this when I used the example of New Year's resolution to describe these two components of Switching Costs in the previous chapter.

For the average construction operative, usable sustainability information is not always easy to come by as Pat Dowling, a UCATT convenor, told me. Without accurate and reliable information, it is not surprising that people will think doing sustainability is a hard topic.[12] If the information we receive is not in the right context, that just makes the Switching Costs so much higher.

One way to cut down the Switching Costs is by allowing people to participate in the decision making. This is the theme of engaging people to take part in suggesting improvements to the way they work (see Figure 16.8). As we have discussed earlier, even this does not remove all the Switching Costs because people can underestimate the actual requirements of 'doing', but it can help cut down somewhat.

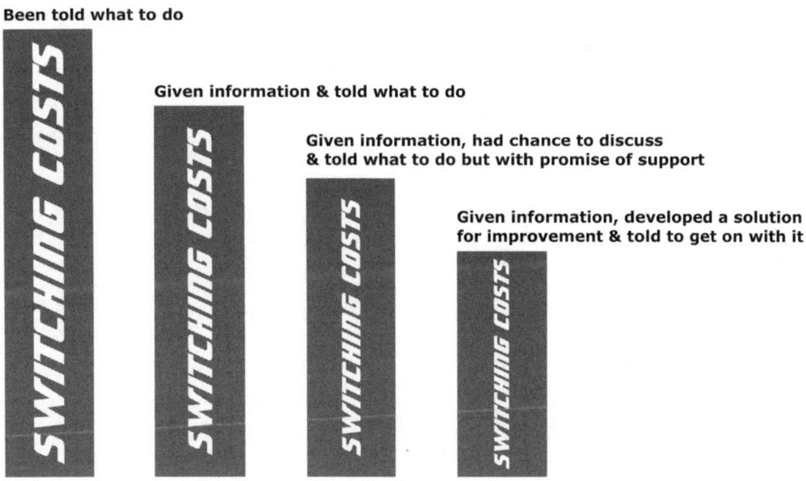

Figure 16.8 Reducing the switching costs

12 Trying asking the average construction sector professional to calculate the carbon footprint of laying a row of bricks. Is it simple? Not really.

Another way of dealing with the Switching Cost is to remove the 'business as usual' option. This means everybody has to accept the Switching Cost because the alternative is no longer available. We have moved to the new world described at the beginning of this book and we need to look at the Switch Costs as an inevitability of work.

Last word on respect – so far all the discussions about the 3R has been about implementation, but respect goes back a long way before innovation. I am always amazed that many supervisors and managers feel that there is no need for their team to know about basic sustainability information, such as the amounts of energy used to make cement, for example. In the NoWaste workshops I ran at Stratford, operators are often astonished to find out that our industry wastes about 100 M tonnes of materials or that nearly 19 per cent of all materials arriving on site end up in a skip ('every fifth pallet!'). Why do we think that our colleagues do not need to know? When pressed, the answers I get include: 'they won't be interested'; 'they are here to do as they're told'; or 'they don't care'. Wrong on all three counts!

Construction workers are interested, and they do care because just like everyone else above them in the construction hierarchy, they have a professional pride in what they do. By keeping their workers in the dark, the managers are depriving themselves of a most valuable resource – the ability to solve problems and innovate. Without useful information, how can we make good decisions? If we cannot make good decisions, then the result is an industry that can only compete on costs rather than win on innovation.

17

Transforming The Organisation

Transforming the organisation is a major task and made harder by the recession, the credit crunch and the slow recovery since 2008. However, the 'backs to the wall' state of mind also makes change easier because the softer options and choices are rapidly disappearing.

Throughout this book, we talked about sustainability as a transformation activity; so how does it happen? Do we hire consultants and let them get on with it? (Yes! Yes!) Do we buy a training programme and get everyone NVQs and trust that to bring us to the promised land? Or do we think about what we want to do as an organisation and then just get on with it ourselves?

Like a lot of things in life, there are bits of answers in each solution. We need to make sure our own people are engaged and on board, we may need some consultancy support to make sure it goes smoothly and we will need an integrated training programme to bring everyone to a point where the whole company can move forward to make sustainability happen.

Transforming your organisation (see Figure 17.1) from a traditional way of doing business to the 'sustainable' way of doing business means taking care of several different things:

1. **Develop a suitable strategy and policy framework**: this is where it all starts for the organisation.
2. **Making sure the formal obligations are met**: this includes compliance with legislation (however onerous) as well as making sure other formal obligations are discharged. At its most basic can be a tick-box-type checklist, so assign someone to make sure all the necessary elements are taken into account. Nevertheless, bills still have to be paid so there may be some constraints on how extravagant the sustainability policy can be (see Chapters 2 and 3 for more details).

3. **Changing people's behaviour**: this is hard to do because it is not a list of tick boxes. People need to be motivated, bribed or even coerced into changing their behaviour (see Chapter 16 for more details).

4. **Managing the transformation process**: an effective package comprising an integrated policy framework, an on-going improvement programme, the Core Intent statement, training and support as well as the 'control' element of DMAIC can help prevent policies being misinterpreted or not followed (see Chapter 15 for more details).

5. **Remember**: transforming the company is like changing yourself (see previous chapter). There is a planning (thinking) element and a implementing (doing) element. Doing the former does not necessarily mean the latter will happen.

6. **Make sure your transformation programme is an integrated across the whole organisation**: Peter Head of Arup, expressed one concern: there may be a danger in that many organisations are looking at sustainability in a piecemeal fashion rather than in a whole or integrated way.

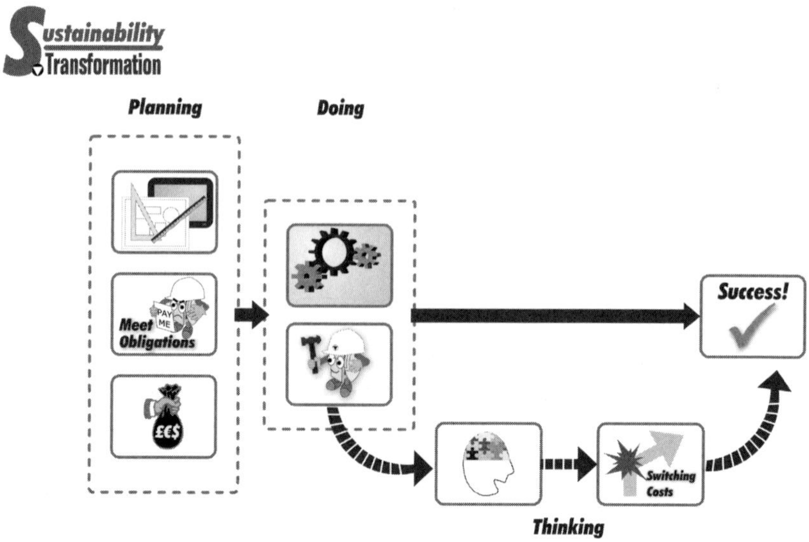

Figure 17.1 Transforming your organisation

Establishing a Policy Framework

First, we need to decide what we want our sustainability policy to do? After all, is not sustainability just common sense? Would not a simple Core Intent statement plus some Tool Box Talks do the job? Furthermore, can we not just respond to events as they come? Since we have decided that sustainability share a number of similarities with Lean operations, so why do we not just stick to a Lean policy and address sustainability issues as part of legislation compliance, cost management or whatever when issues arise? This approach is certainly possible provided that the business environment is static and stable.

However, whether in good times or bad, construction is rarely static or stable, there are always new regulations, new designs, new materials and new finance or contract mechanisms to keep us alert. We have already said that sustainability is both the new 'black' and the 600 pound gorilla. With something that is fashionable, un-ignorable and changing, plus an economy trying to get out of the pits, we will need a specific strategy to get through this period.

Deciding what you need to do in a policy is an important activity at whatever policy-making level in any organisation. The exercise is necessary because it allows your executive team to come to consensus about a policy (and hopefully, consensus on the need to do something), and also to focus the collective minds on what needs to be addressed in the policy.

What Are Sustainability Policies For?

This may seem like a redundant question since 'everyone' knows what a sustainability policy is for. Yes, over one million definitions of what it is for in our industry. Policies form a framework of directions for a company and guide the way it functions.

A sustainability policy framework can do many things, including:

- Acting as a wish list, such as the corporate mission statements on sustainability.
- Serving as a foundation for a strategy to navigate the future directions the organisation takes to fulfil its sustainability mission.

- Becoming the basis to shape positive behaviour.
- A tool to clarify the way we work through codes of practice.
- A means to communicate the wishes of the organisation.

A sustainability policy framework, therefore, can help us clarify what it is we wanted and how to deliver all these (and more). It can also allow the organisation to guide itself through its sustainability activities: to safeguard the community; to protect the environment; and to deliver financial value. It enables the organisation to decide on a sustainable purchasing and reuse strategy. It encourages the engagement of the various teams on a project towards improved safety and creates a framework to update Method Statements. A sustainability policy also means that the company can deliver a message of what it wants to achieve in a structured and managed way.

Although a policy for sustainability is often a policy for common sense, it is not always easy to come up with something that is financially sound, treats people with respect and minimise the impacts on the environment? We can all look at the examples we see around us from government policy downwards to know that the best intentions do not always deliver the desired outcomes. Is formulating an effective strategy and policy framework that difficult? Do you really need an MBA degree or expensive consultants to help out? (Yes to the latter, of course!)

In this section, I want to show you how you can develop a realistic strategy and policy framework that can help your organisation, your teams and partners to work towards sustainability goals using the House of Sustainability, a simple but detailed and effective technique.[1]

The House of Sustainability (HoS)

This approach was adapted from the House of Quality,[2] originally developed to create new products and services. The HoS (see Figure 17.2) can be seen as a vastly expanded What/How Matrix – in this case about seven matrices. All the matrices have a similar relationship of the 'what' being the need and the 'how'

1 'Simple' is relative, the House of Sustainability is simpler than the full House of Quality.
2 The House of Quality or Quality Function Deployment is a quality management tool. In its original guise, there can be up to 17 matrices.

being the response. The HoS for construction can be used to develop products and services, policy and activities.

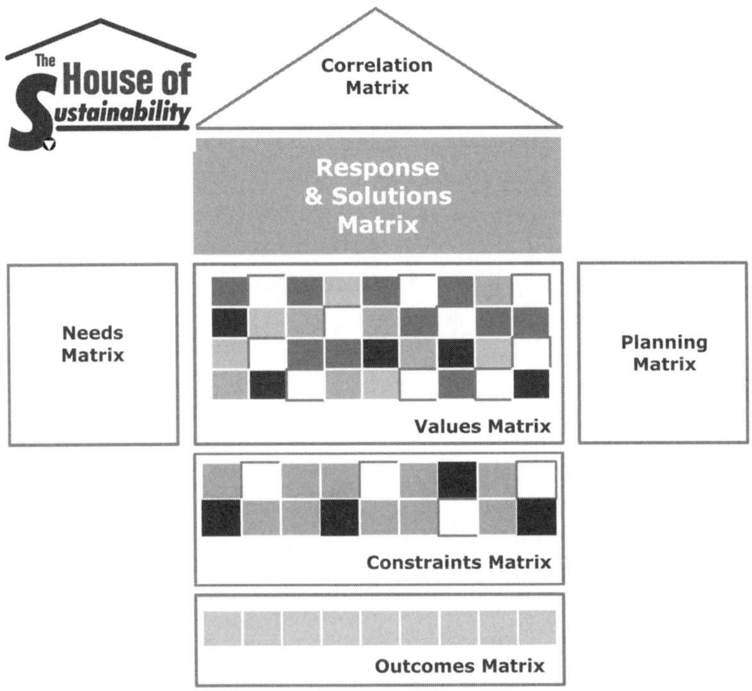

Figure 17.2 The House of Sustainability

The seven matrices are:[3]

1. Needs Matrix
2. Planning Matrix
3. Responses and Solutions Matrix
4. Values Matrix
5. Correlations Matrix

3 'Seven matrices' – although some of these do include sub-matrices ... but who's counting!

6. Constraints Matrix

7. Outcomes Matrix

1. The Needs Matrix

The Needs Matrix is about client expectations – what they wanted your company to deliver. This can be high-level expectations, such as a building that fits with the strategic needs of the client, or low-level requirements, such as the type of picture for a photo concrete wall.[4] These expectations can be from a framework agreement or the tender specifications, or even verbally from a client (*'it would be nice to have …'*).

This list of needs is not only the client expectations, but also includes society demands (directly or indirectly) as well as legislation. To determine this lists, companies often carry out surveys of both client views and social trends.

Although legislation is not part of the client's needs and instead a constraint, it is included in this matrix because the impacts can be very profound and there is little point carrying out a HoS exercise and then finding the proposed way forward will clash with legislation. Therefore the 'rules' are treated as embedded elements of the client's needs.

Once you have gathered the necessarily information for such a list, you will need to sort these needs into some order so that you can prioritise and also to combine or remove if required.

Some of the sorting methods have been described earlier in this book – voice of the customer, affinity diagrams and so on. Before any sorting can take place, however, you should get some clarification from your clients. Is the needs list a wish list only? Or are these items something they are prepared to pay for? This clarification is necessarily for both strategic and project-level sustainability policies (see Figure 17.3).

4 This is one of those wonder technologies that allows a photo be reproduced onto a concrete surface. Hering International seems to be the experts in this.

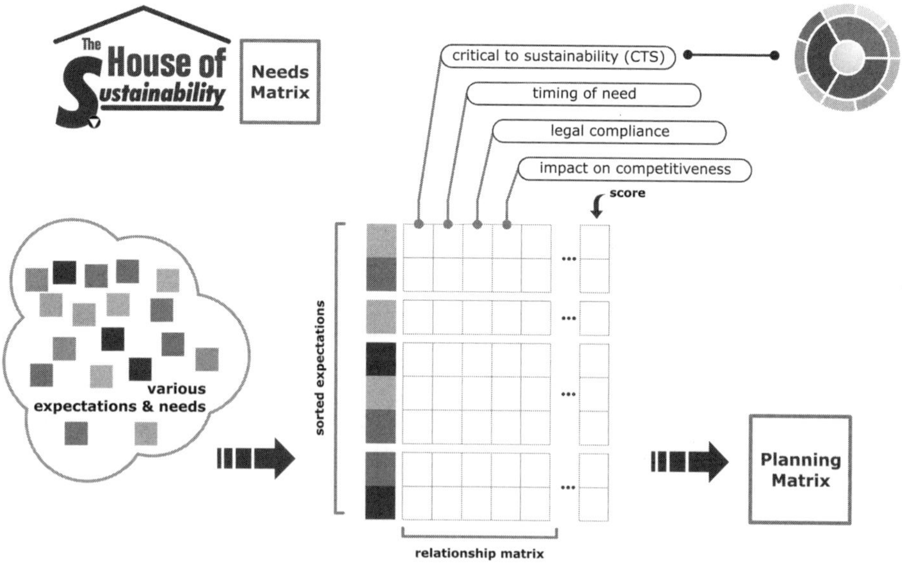

Figure 17.3 Needs Matrix 1

CLARIFYING STRATEGIC NEEDS

The importance of clarifying the client's needs was explored earlier in Chapter 15. However, for developing a strategic framework, the needs analysis has to be more detailed than that for a single project. Specifications for a building essentially define the client's need for a specific project and the construction industry is familiar with how to deal with those. When a company is developing a strategic sustainability policy framework, then there is the need to go wider than just clients. Future legislation, social trends, skills availability as well as new technologies, all play a role. This is when partnerships with clients and subcontractors can help define the future needs of the industry. The trade media as well as universities both have capabilities for crystal ball gazing and a company should take these views into account.

SORTING AND CLASSIFYING THE NEEDS ITEMS

A simple way to sort these needs is to give each a score or rank according to their importance. For example:

5: critical/safety/legal need – must be satisfied

4: very important

3: highly desirable

2: nice to have

1: only if it is very easy to do.

Although the scoring example above[5] is very simple, it is probably adequate to sort out the 'wish list' items from the 'must have' items. However, for sustainability, we should also take into account the 'Critical to Sustainability' (CTS) aspect of any client need. A simple CTS analysis (first explored in Chapter 1) can be carried out using the model. Each wish list item can then be assessed for their CTS value – how much they score on each of the nine elements of sustainability (see Figure 17.4).

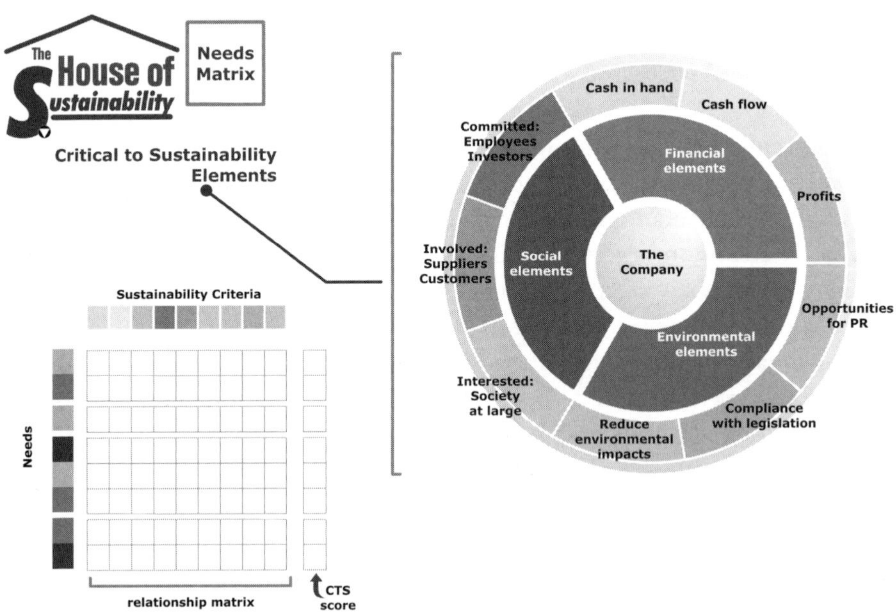

Figure 17.4 Needs Matrix 2

5 This is also known as the Likert Scale.

Now you have two lists: one comprising client importance; and the other, a criticality analysis. The reason for all these classifications and sorting is that with all the information on views and opinions, and the subsequent classification and sorts, the needs analysis will possibly resemble a 2,000-piece jigsaw puzzle. You neither have the time to put all 2,000 pieces together nor do you necessarily have interest in some of the background pieces. The point[6] is that you want to see the big picture generally and you probably also like to know every necessary detail that make up the big picture. (Well, perhaps you do, but you are unlikely to have the resources for that.)

The sustainability specialist can consider adding some weighting elements to these various declared need. A simple weighting system is to ask the following questions:

- How critical is this to sustainability (CTS)?
- How soon is this need likely to require fulfilment?
- Is there a legal compliance requirement?
- What is the impact of fulfilling this need on competitiveness?

Although the simple weighting system above is equally suitable for strategic sustainability issues as it is for project-level sustainability issues. You may want to use a different set of questions more suited to your company's situation. As long as you are consistent in asking your questions, it should not affect the accuracy of the analysis.

The end point of the needs analysis is a list of the top-scoring need items. These may then be grouped into financial, social or environmental categories, or others of your choice. This list of needs becomes the Needs Matrix, a finalised list of expectations – the 'what' list.

Earlier, I explored the need for clarifying the needs of various groups. The clarification should also include whether the various corporate sustainability missions are compatible with each other. In any construction project, there are a number of different companies working together to deliver a common goal. But each of these companies, be they client, designer, cost consultants,

6 This is metaphorical – people do jigsaw puzzles to put all the pieces together, not to get a sense of the picture, which is what we are after.

project managers, builders, specialists, is likely to have a somewhat different interpretation of how sustainability is achieved in their company. The way to ensure that these are compatible is to carry out a correlation analysis – the method is exactly the same as the Correlation Matrix, described later.

2. The Planning Matrix

This matrix is about the current performance of your company and how the available capabilities match up to the needs (see Figure 17.5). The items in the Needs Matrix now becomes the specification for our products and services. The Planning Matrix can have several elements: a current state analysis as well as a future state analysis. It can also include just about any measure that you think are relevant in addressing client needs. It is possible to use the outcomes of the Planning Matrix as weighting factors in the Constraints Matrix, especially if there are factors which may indicate that a particular issue is very hard to do or critical to sustainability.

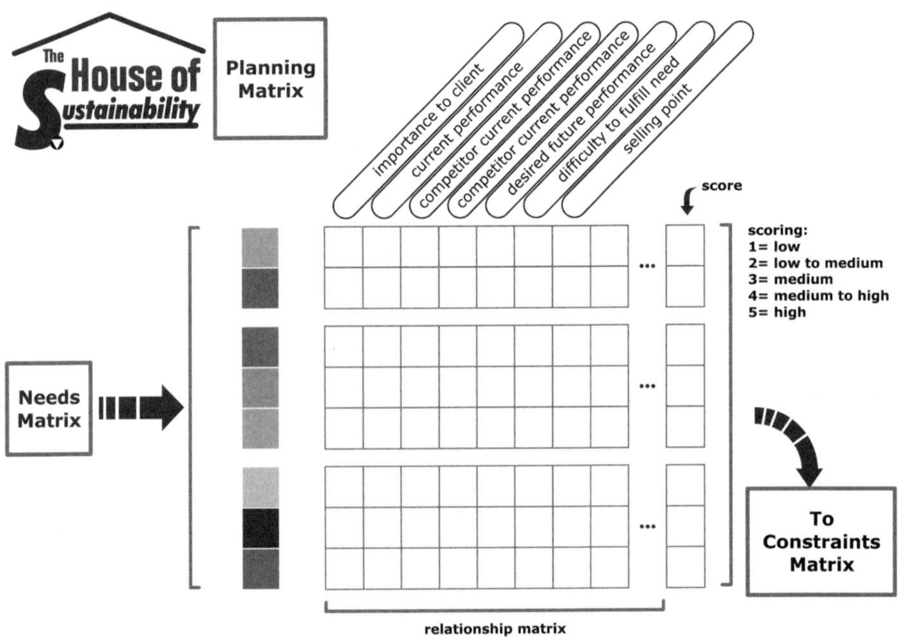

Figure 17.5 Planning Matrix 1

In general, the Planning Matrix should always include:

- your current performance in meeting these needs;
- your competitor(s)' performance in meeting these needs currently – this may be an estimate or an observation perhaps via a third party;
- the level of performance you need to reach in future to stay competitive;
- easiness or difficulty of achieving the desired level of performance;
- selling point: does fulfilling this need really help our sales effort?

While the first three elements are fairly straightforward although you would need to estimate your competitor's performance, the final two elements are more critical.

Your performance and your competitor's performance give you a feel for the current state. What is the client experience of your performance? How do your competitors perform in the same areas? By matching your performance towards the various needs against your various competitors, you will be able to see who has the better performance for each need. It also allows you to explore what level of performance you will need to achieve in order to have a competitive edge. Often this means you need to raise your performance to a higher level.

This part of the matrix offers additional details of analysis (see Figure 17.6), you may need to ask both clients and suppliers to rate your performance against your competitors as well as having your own colleagues giving their best estimates (or guesses).

The hard part is to set the level of future performance. You need to consider what level is expected:

- What is the expected improvement from your competitors in this area? Do you therefore need to be: 25 per cent better? 10 per cent better? 5 per cent better? As good as? Nearly as good as?

- What will future client expectations be? 25 per cent better? 10 per cent better? 5 per cent better? Stay the same?

- What about society demands? Or legislative targets?

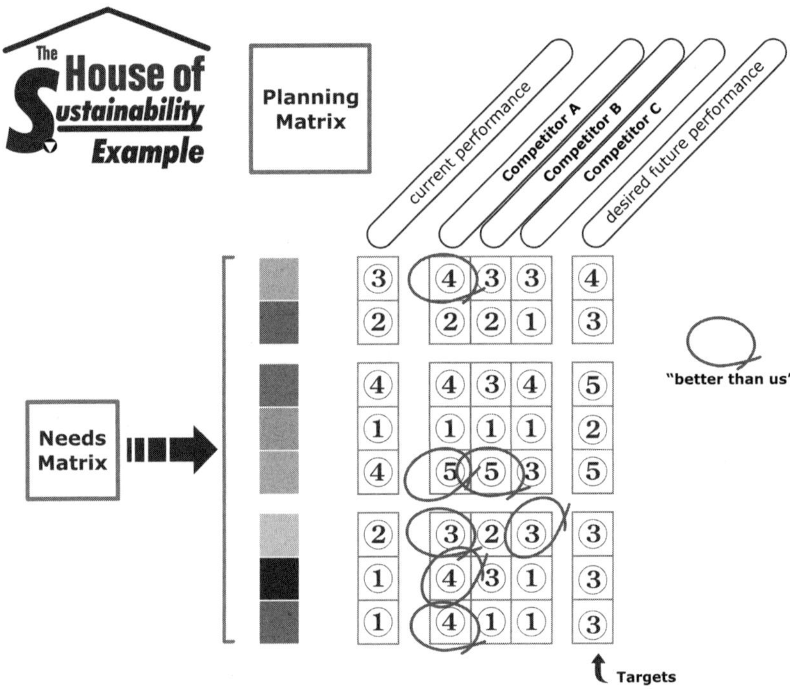

Figure 17.6 Planning Matrix 2

This analysis is critical because it determines what level of resources you will allocate in the future. Over-setting the sustainability performance targets is good for the world generally and earn you extra kudos. But if your client is expecting a gold-plated service, then delivering a solid gold service will make them happy, but it will not necessarily increase your income. A solid gold service may also costs more in terms of people time and money as well. The extra kudos is not likely to buy you much if you end up not being able to meet your financial obligations. Under-setting your performance targets will mean you fall behind your competitors and lose both income and prestige.

DIFFICULTIES AND SELLING POINT

There are two aspects of difficulty to examine: the difficulty of upping performance generally and the difficulty of doing a specific task. The jump from whatever your level of current performance to future performance brings several points to consider:

1. **Same activity**: raising performance by one level is not necessary similarly difficult; going from level 1 to 2 is often easier than going from level 4 to 5 since the base line (at level 1) is much lower.

2. **New activity**: raising performance through a new activity brings with it an entirely different set of difficulties. It is not just about doing better, but instead it is about doing something else and also doing it well. These, however, may lead to greater risks and costs.

The actual difficulty in fulfilling the needs is different to carrying out the improvements. A difficult task is a difficult task: designing the master plan for the London 2012 Games is a hard task in itself. But knowing that you got to raise your performance to a level above that of, say, Arup, for example, brings an entirely different definition of difficulty.

It is therefore important to link these difficulties with the selling point. If a particular need is important to the client, then it will sell. However, experience tells us that in construction, a wish list during the design stage may not necessarily end up in the building. There is often a stage called Value Engineering to go through between design and reality. To some, this is basically a cost-cutting exercise but to others, this is an opportunity to remove frivolous items, or items that do not become a selling point.

The selling point, therefore, needs to be matched to the Critical to Sustainability Analysis. The various needs will still have to deliver the CTS values whether the Value Engineering process reduces their scope.

The final score of the Planning Matrix allows you to rank the Needs Matrix elements. As most companies have limited resources, the lowest scoring needs can be placed as lower priority, perhaps for the next set of policies. These scores represents factors that will have a bearing on how your policies will be shaped and how the responses can be implemented. The most important element should then be placed into the Constraints Matrix where they can be measured directly against the responses.

3. The Response and Solution Matrix

This matrix is the technical response[7] to fulfilling the needs. It is the 'how' component of the What/How Matrix. This is one of the simpler matrices in the HoS analysis in that you list the activities you reckon will deliver the needs. Essentially you prepare a list of possible activities that may fulfil one or more need element.

There is, however, some sophistication to this matrix shown in Figure 17.7.

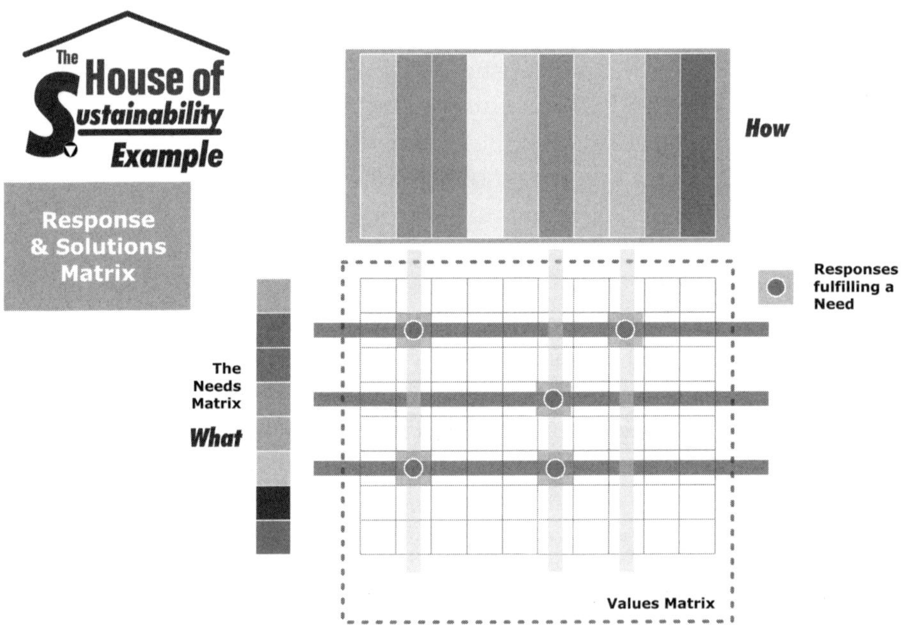

Figure 17.7 Response and Solutions Matrix 1

7 If you think 'technical response' is a bit of jargon, consider this – House of Quality specialists call this 'substitute quality characteristics'! And apparently the client's needs are called Quality Characteristics – so there you go!

Abstraction

The more abstract a technical response, the more likely it will fulfil more than one need. At the working level, abstraction like this is probably less useful than at the strategic level. However, the need element is likely to be a lot more abstract (or vague) than the technical responses. Nevertheless, the abstraction can be readily reduced by cascading the what/how relationship to the next lower level. The how from the first round becomes the what of the second round (see Figure 17.8).

For example: the top-level need ($What_1$) is 'engage operatives on site', a well meaning if vague expectation. At the strategic level, this is perfectly acceptable as the technical response (How_1) can be 'respect, recognition and reward'. But at the project level, this How_1 becomes the new $What_2$: 'what do we have to do to show respect?'. The technical response at this level, How_2 can then be: 'by encouraging every supervisor and manger to say thanks when someone comes up with a waste reduction idea'. $What_3$ then becomes: 'what do we need to do to encourage this?' and so on.

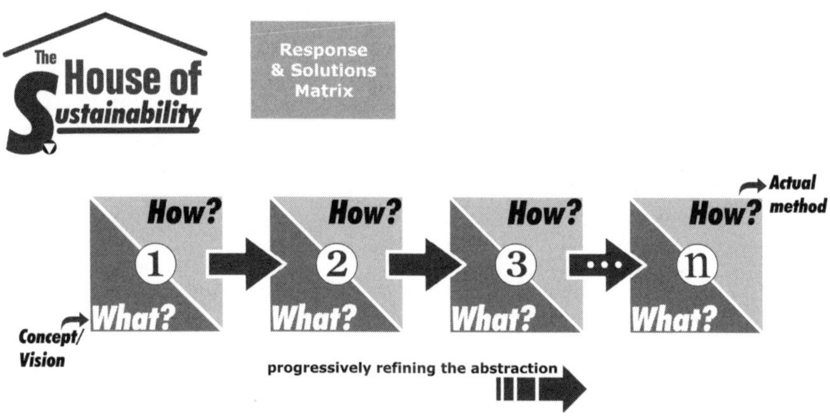

Figure 17.8 Response and Solutions Matrix 2

The increasing concreteness of the what/how analysis can eventually lead to an action method or activity at stage How_n. This allows you to work out the actual resource requirements: time; effort; and cost. Although at this stage, you may only need a rough estimate of the time, effort and cost. Once the technical response is selected as the way forward, then a detail work breakdown analysis will be necessary.[8]

Performance level

In the Planning Matrix, the future performance is explored. In this matrix, the type of performance required can be defined: more is better; or less is better; or on-target is best.

1. **More is better**: this parameter is for situations, such as recycling rates, satisfaction rates, numbers of innovations and so on. The point to watch for is that if more is better, then infinity is surely best. This type of thinking can be expensive to implement and hard to do.

2. **Less is better**: this applies to accidents; waste volume; defects and errors; environmental mishaps; and so on. Again we need to watch for unrealistic targets of zero defect, zero accidents and so on. While these are desirable, they may bring about other impacts on the project, such as excessive time and costs.

3. **Target is best**: this covers areas, such as meeting building specifications and component specifications. A simple example is making sure lifts stop on the right floor and not higher or lower! This also covers contractual issues and other obligations, such as paying the tax.

Ambiguity comes in with cost and schedule – although target cost is good, a bit less than the target is better as it soothes the client. However, too much below the target can lead to errors and profitability issues – the record of underbidding in our industry is a good example of the less-is-better mentality getting out of hand.[9]

8 This cascading is also known as the Sustainability Function Deployment, a bit of mega jargon. In the quality management world, it is called the Quality Function Deployment. However, unlike the quality people, I do propose a health warning: going down to the individual task level can be very laborious and time consuming, and you may also need very large sheets of paper.

9 *Building Magazine* (issue 28, 2009) reported on a case of a PFI hospital programme where initial underbidding led to a host of difficulties.

Performance measures

A technical response will meet a specific clients need, but we still require some type of performance measure to check that the need is met. The performance measure (see Figure 17.9) can be found through a cascading of the What/How Matrix to a lower level where technical response becomes a method or an activity (level How_n). With a defined method or an activity, it is then possible to match this to the specific client need and determine the required performance measure. This can then be further matched to the more, less and on-target criteria.

The performance requirements are also developed from the information gathered during the Planning Matrix analysis. The future performance and selling point analyses provide the necessary information in assessing performance requirements.

This point was discussed earlier as one of the Eight Hidden Wastes: over doing.

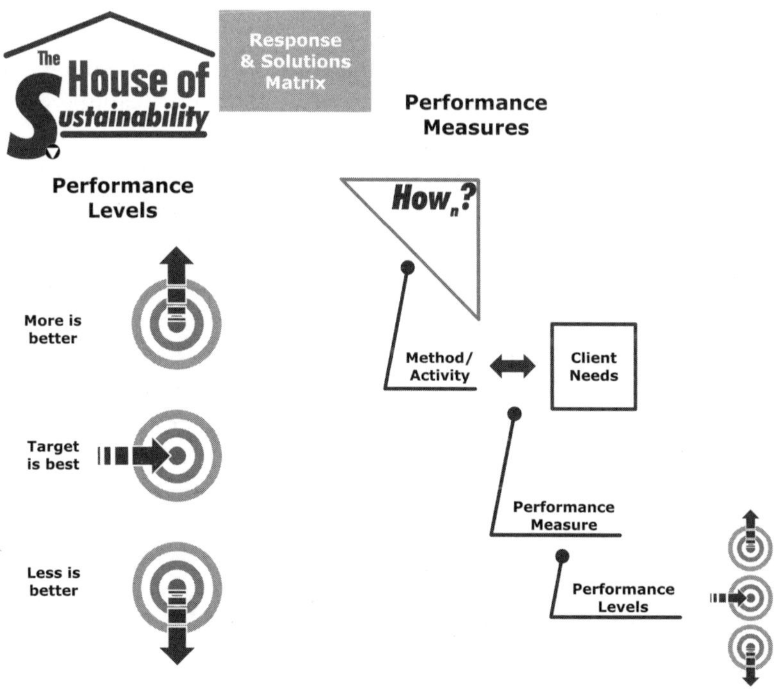

Figure 17.9 Performance measures

4. The Values Matrix

This is simply the relationships between the Needs Matrix and the Responses and Solutions Matrix. The relationships can be one-to-one, one-to-many or many-to-many. A single technical response may be able to meet one or several needs (see Figure 17.7).

For example, a set of sustainability needs may include: lower running costs; a happier workforce; and reduced environmental incidences.

> **Technical response #1** may be an across the board 5 per cent cost-cutting: this will meet the lower running costs need but unlikely to have a lot of impact on a happier workforce and reduced environmental incidences.
>
> **Technical response #2** may be a Gain-Share programme: this will help reduce costs through innovations, the workforce gets happier from the share elements and environmental incidents reduced through the innovations.
>
> **Technical response #3** may be a workplace effectiveness programme: the costs go down because of reduced waste levels and the workforce is happier because they are working in a safer, cleaner and tidier workplace. But the environmental incidents level may not be particularly affected.

Although the relationships can be represented in an abstract way by having ticks or nothing, a ranking system is recommended[10] because it allows a quick assessment of how the technical responses impact on the needs and how the needs are met by the technical responses (see Figure 17.10).

A simple scoring system, the 1–5 scale, for example, allows you to check quickly which need requires a more detailed response. Adding these scores for each response quickly shows which offers the best value for money in terms of its overall impact. It also lets you review whether you have to come up with further ideas if the first group of technical responses does not fully deal with all the needs (see Figure 17.11).

---

10 Although reference books on the House of Quality generally shows an array of symbols to represent intensity, my pragmatic suggestion is: use numbers – there is no need to remember any symbols and it is easier to produce either on paper or spreadsheet.

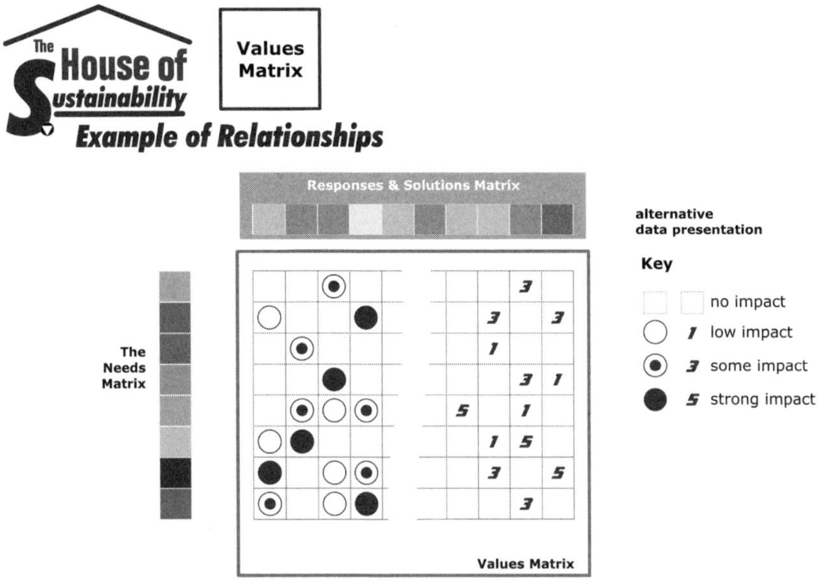

Figure 17.10 Values Matrix: examples of relationships

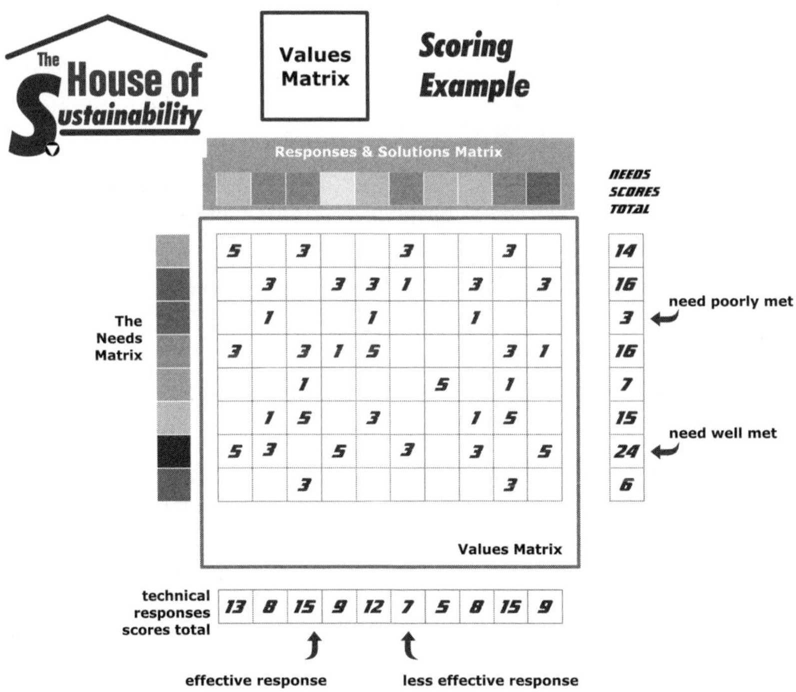

Figure 17.11 Values Matrix: scoring example

5. The Correlation Matrix

The roof-like shape[11] of this matrix gives rise to the name House of Sustainability. The Correlation Matrix is a simple tool to make sure that your various technical responses either work well together or do not interfere with each other (see Figure 17.12).

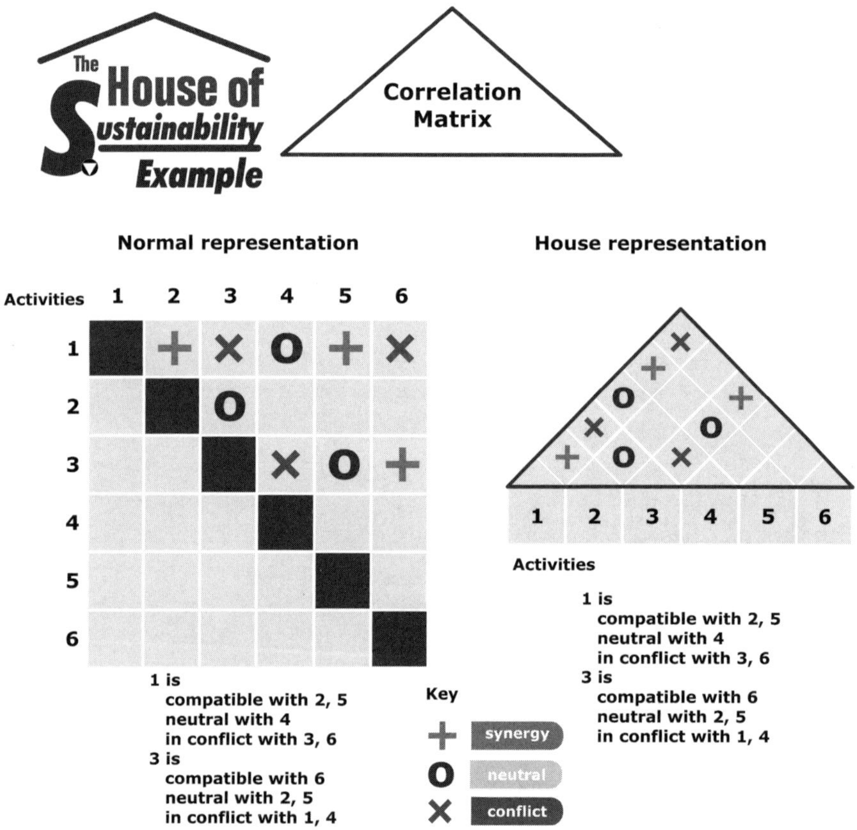

Figure 17.12 The Correlations Matrix

The relationship can be as simple as the three listed below:

1. **Synergy**: high quality and low waste.

11 The design of the roof shape also makes the matrix a bit difficult to understand quickly. But as we know well in construction, sometime a good design trumps usability!

2. **Neutral**: good crane scheduling and secure bicycle sheds.

3. **Conflict**: a lowest bid price programme is unlikely to work well with best social responsibility components.

The correlation range can be expanded to provide intermediate points between the three above. However, a 'likely to conflict' score is not a lot better than 'will conflict' and you will probably need to manage it like a 'will conflict' case and nothing like managing a 'neutral' relationship.

The basic reason for the Correlation Matrix is to spot the conflicting responses and prepare for them either by:

- Coming up with new conflict-free technical responses.

- Going back to the drawing board with the potential conflicts and deal with them.

- Live with the choice and carry out risk management planning.

After this analysis, the sustainability specialist will have a set of solutions that will meet the client's needs whether at the strategic or the project level. Furthermore, these solutions take into account the current performance, expected performance, sustainability criticality[12] and whether it can contribute to sales. What the analyses so far have not taken into account are constraints.

6. The Constraints Matrix

Whereas the Correlation Matrix focuses on meeting client needs, the Constraints Matrix instead looks at the issues within a company and how these can impact on delivering value to clients (see Figure 17.13).

Other than the usual list of: can't afford it; won't afford it; and we've got better things to do, what are the constraints facing the sustainability specialist? Perhaps a lack of skills or a lack of determination?[13] The wrong time? A different

12 This is a mouthful to pronounce, perhaps by the next edition, the jargon will have evolved to a smoother phrase.

13 A senior UK manager did reveal to me once that 'proper' Lean construction is a bit too hard for his company and he would prefer a dumb-downed version.

corporate agenda? There are all kinds of reasons why the sustainability initiatives from the Response and Solutions Matrix may be hampered or constrained.

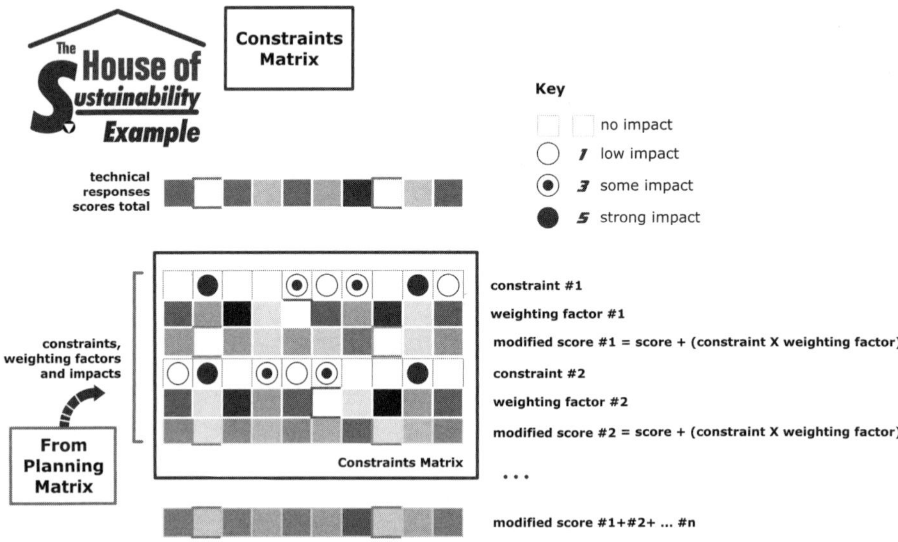

Figure 17.13 The Constraints Matrix

Of course, the constraints are not always negative like the list above. There are positive constraints too, for example:

- A corporate emphasis on work–life balance may mean the teams cannot burn the midnight oil (so no 'away day' of 24/7 working to develop a full House of Sustainability analysis over the weekend!).
- The corporate mission statement and the Core Intent statement may channel the focus in one particular direction (a Core Intent statement of 'health and safety is really important in this company' will shift emphasis).

Other constraints can include the shortage of suitable resources, geographical issues, such as city centre building with limited space for on-site waste segregation or requirements for quiet hours for the neighbours.

Using the scoring from the Values Matrix, it is possible to check how the values delivered by the responses can be shifted by various constraints. The subtleties of the Constraints Matrix is that the individual constraints can have weighting factors. So it can include various corporate emphasis, such as the Core Intent statement or a corporate aim to gain a Business in the Environment gold-level award, for example.

Using the Core Intent statement example of 'health and safety is really important in this company' as a constraint, then the relationship of this constraint on the various technical responses will vary from high for 'updating method statements' to low for 'encouraging people to eat more fresh fruit'. Then the importance of the constraint is taken into account through a weighting factor decided by the sustainability specialist after consultation with colleagues.

The final modified score describes how the various technical responses will deliver client needs while taking into account constraints in the company.

7. The Outcomes Matrix

This is the final matrix in the House of Sustainability and its function is to clarify the analysis and produce a prioritised set of sustainability solutions.[14] The final scores from the Constraints Matrix are ranked simply and it is then possible to show how well each solution will perform (see Figure 17.14).

The Pareto Analysis or the Lever Rule are fancy names for the theory that 80 per cent of impacts come from 20 per cent of the activities. Once the final scores are ranked, you can then determine which solutions should be implemented immediately, and others when there is resource available. The Outcomes Matrix can contribute in deciding which batch of solutions should be implemented together for balance purposes or to maximise synergy. It also allows solution implementations to be staged to avoid conflict or minimise costs.

14 The technical responses can be called solutions now since they are ready for deployment.

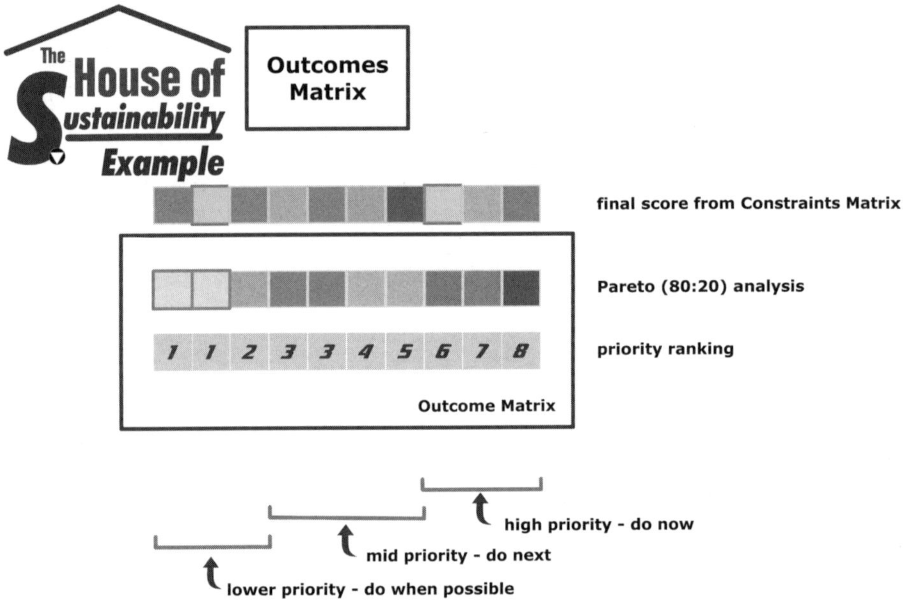

Figure 17.14 The Outcomes Matrix

The House of Sustainability is a simple tool in concept but allows a complex analysis and it will involve a significant amount of time and effort. However, it is possible to use this tool to develop a coherent set of sustainability policies for the company as part of the transformation from a traditional way of doing business to a sustainable way of doing business. It is a scalable process and can be used for strategic and operational applications.

18

Was Egan Right?: Thoughts on Sustainability in Construction

In 1998 Sir John Egan, then the boss of BAA,[1] produced the Rethinking Construction Report[2] which advocated a range of new and good practices for the industry. Ten years later, in a speech in Parliament,[3] Sir John awarded the construction industry four out of ten for trying. Why is this Report important and why did our industry score so poorly?

The Egan Report was about a lot of things, but mostly about how to make the industry perform better. There were five key drivers for change:

1. committed leadership;
2. focus on the customer (both the client and the end-users);
3. integrated processes and teams;
4. a quality-driven agenda; and
5. a commitment to people.

There are several reasons why the Egan Report is relevant to sustainability in construction. The obvious ones are the five key drivers above. Each of the five drivers is also key for sustainability. Although Egan did not explicitly mention

---

1 Before running the British Airports Authority (BAA), Sir John used to run Jaguar cars and he was running a water company in 2009.

2 The Rethinking Construction Report is available from the UK Department for Business, Innovation and Skills website. All subsequent references of the Report are from this document.

3 As reported in *Building Magazine*, 21 May 2008.

sustainability,[4] the underlying themes and drivers are compatible with the sustainability objectives: quality; cost reduction; and respect for people.

The less obvious reason is that, to me, the Egan Report is a thinly disguised call for the construction industry to adopt Lean practices. Sir John saw Lean construction as the way forward for sustained success in the industry. Since I am also suggesting that Lean operations is a realistic way forward for sustainability, I guess that makes me an Egan advocate.

So what happened to cause Sir John to offer a four out of ten score after ten years? The Rethinking Construction Report was ambitious and in some areas, unrealistically so. There were insufficient incentive for change and the construction industry was not (and is not) the car industry. Others have expressed the view that ten years is insufficient for changes of such scale and scope.[5] It may be that sustainability is a 'generational' event, as suggested by Murray Coleman. There is also little government leadership in really pushing for the Egan agenda. Instead, as Peter Rogers of Stanhope, a developer, said to me, 'people had to push the government to make things happen because individuals are limited in what they can do on their own'. What happened was the government coming in with its lowest bidder wins attitude (never officially acknowledged, of course) that encourages just about everything the Egan Report is against. These and many other factors all hampered the successful adoption of this pioneering Report.

These obstacles start with Egan's suggestion of a commitment at the leadership level towards continuous improvement. This is hard to do. It is hard because the UK construction sector is a closed market with a fairly level playing field. This meant that new ideas are not particularly easy to introduce. For example, when a foreign construction company buy an existing UK business and gives it a new name, the core is still a UK construction company. Besides, the construction takes place under UK laws, with UK subcontractors, mostly staffed with UK people with suppliers that abide with UK laws and practices. So where is the new thinking going to come from?

As long as everyone makes a relatively decent living, what is the incentive to change? Egan complained about the low level of profitability in construction,

4 Not surprising especially since the Kyoto Protocol came out in December 1997, more or less at the same time as the Egan Report.

5 Some were even less complimentary, one colleague from a major organisation offered the view that 'Egan is a madman'.

but if that is really the case then why are there 163,000 companies (as reported for 1998) in the industry then? If companies can get by with business as usual, then even if the wolves take a few weak ones every year, so what? As long it is not your company that goes under, there is little incentive to change. The UK business culture is not a long-termed one and asking a very conservative sector to lead the change may be an unrealistic hope.

Egan also mentioned that he first thought the government will place his Report in the rubbish bin. Other than setting up Constructing Excellence and assorted demonstration programmes, the government has not showed much commitment to Report with its revolving door policy for construction ministers[6] or its lowest bidder mentality.

The focus on customers is an interesting one. In construction jargon, the customer tends to be called the client. Egan calls for more attention to be paid to end-users and complained that too much of a subcontractor's focus is on their clients – the next higher level of 'employers' within the hierarchy of subcontracting. This has not changed a lot over the years for good reasons – it makes sense to pay attention to the people paying your invoices! This practical but short-range vision means that the focus is invariable on the immediate job and not on the bigger picture, which leads to the next key driver for change: integrated teams. David Farebrother of Land Securities, a developer, said that he believed that end-users and developers should work together to reduce energy use and in his view, the long-term vision shown in these areas by Land Securities was attracting more clients.

In car making, integrated teams tend to be the way forward. Just-in-time deliveries and other Lean practices mean that suppliers and customers are linked closely where subcontractors are encouraged to share ideas and savings. This was Egan's vision for construction. The result was a number of frameworks being set up where contractors and suppliers build extended relationships. These relationships are supposed to enhance best practice and bring about the Egan vision. In 1998, BAA, Egan's company was a pioneer in frameworks. In 2009, however, BAA is expressing a preference for lowest bids and an end to frameworks.[7] The world has changed and ideas move on. Nevertheless, integration is still seen by many of the people who really care about our industry as the right way forward. Jon de Souza, Claudine Blamey,

6 Of course, I may have got this all wrong. It may be because it is so popular that the ministerial appointment had to be shared on an 'everyone gets a turn' basis – what do you think?

7 *Building Magazine*, issue 25, 2009.

David Farebrother, Phil Henry and others, I talked to, all agreed that there should be more not less integration.

Egan also suggested that the construction industry should go for more generic products which in car manufacturing would mean common platforms, shared components and so on. In construction, generic products tend to bring up the image of a plastic pod. Furthermore, the value stream in car making is dominated by the assemblers – the car makers, with designers, suppliers mostly playing second fiddle. Regulators and other bodies in car making are there for legislative compliance, not style compliance or good taste.[8] In construction, we have very influential superstar architects, superstars (see next paragraph) and powerful local authorities who can all block developers.

Despite Egan suggesting more power to the clients, the construction sector can be swayed by forces that have little financial interest in the project. The furore of the Chelsea Barracks redevelopment programme in Central London during 2009 is a good example of how our industry works sometimes. A developer hired a superstar architect to design a set of building which the style gurus approved. The local residents did not agree since it changed the way their neighbourhood looked and engaged the support of a superstar amateur (well, Prince Charles) to lobby the developer. End result: development shelved, a lot of grief for the architect, egg on the face of the style gurus and probably an unhappy client.[9]

Egan also advocated the use of Value Management[10] (or Value Engineering) as a way to reduce waste from a design. Architects do not often like Value Management because it changed their vision. End-users may not like Value Management because the perceived quality of the product has dropped ('*using cheaper methods and materials*'). In Lean manufacturing, the wastage is eliminated by better design and operational practices. The value stream in manufacturing is also often parcelled out to several subcontractors each with different objectives to the client or main contractor. But unlike in construction,

8 This is not a dig at bodies like CABE – the Commission for Architecture and the Built Environment, which is here to 'promote design and architecture to raise the standard of the built environment'. See the CABE website for more. At the time of writing (late 2010), CABE's future is being discussed a part of the UK government's spending cuts.

9 The summer 2009 saga was well reported in the UK general press as well as the construction sector trade press: *Building Magazine and Construction News*.

10 A rigorous process where every item is assessed for their value and lower cost substitutes are introduced if required. A less rigorous but common view sees it as cost-cutting. To some, this means a beautiful design is reduced through cost-cutting to a basic cube with holes for windows.

the power and prestige in manufacturing remain with the assemblers: for example Boeing, HP or Ford. In construction, by contrast, the highest prestige often goes to the design teams, not the builders.

The quality-driven agenda proposed by Egan is one key driver that is not controversial and is relevant to any industry all the time. Egan suggested that construction companies should tap their suppliers for innovations, just like the car companies. He also advocated a sharing of savings from these innovations and after ten years, was firmly on the side of a Gain-Share arrangement. However, our industry is one of the slowest to pay its suppliers, up to 90 days in some instances and sharing does not come naturally either. This type of attitude is not one that encourages suppliers to offer innovations.

The Report also found that many construction accidents were caused by poorly trained operatives working outside the proper procedures. This led to the key driver for commitment to people. It was not just about site amenities but more about supervising and training people to achieve the best. The Report did acknowledge that the fragmented nature of the industry and site conditions did not help.

SO WAS EGAN RIGHT?

Egan was and still is mostly right. He advocated a very exciting (and for some, scary) vision for the construction industry at a time when there was little inclination for change. I mentioned earlier that Lean operations was heavily disguised in the Report, I guess Egan probably thought he pushed it as far as it can go in 1998. Looking at Rethinking Construction Report again in 2009, it is hard not to noticed that a lot of Egan's comments remain valid and many are still current. If we are to extrapolate Rethinking Construction Report into a framework for sustainability in construction, what will we find? Perhaps something that looks a bit like the following.

Rethinking Construction Report for Sustainability – A Possible Extrapolation

To recast the Egan Report in a sustainability context will require the five drivers for change to have their context altered, not changed.

1. Committed Leadership

To make sustainability happen in our industry, we need leadership. Not leadership by talk, but leadership by commitment. Running a construction company is not about public office, so leadership by example is nice but less useful than leadership by commitment.

Often we read about what senior executives or superstar gurus believe the industry should be doing. We also hear of senior executives demonstrating their personal commitment towards sustainability by driving big hybrid limousines or not printing their emails. But this does not make sustainability happen in their companies any more than the government's strategy. To quote a sportswear company, 'Just do it!' It is clearly much nicer to have someone with a personal conviction towards sustainability, but we need to be realistic. If the current generation of leaders in our industry is no more committed to change as Egan's generation, we at least welcomed the pragmatic ones who sees sustainability as a fact of business life. When viewed in such a pragmatic way, we probably should be grateful for the single-minded and determined boss to make it happen than relying on nice people.

Yes, the construction industry was not totally committed to the agenda for improvement Egan proposed in 1998, but now we have environmental legislation to help reinforce corporate commitment. With legislation, commitment matters less than compliance. Therefore even if the current generation of senior managers are no more committed to improvement and efficiency as their 1998 counterparts, environmental management activities now happen because of legislation. Companies will have to spend money to set up their sustainability programmes and this offers the opportunity which the Rethinking Construction Report era did not have.

This leverage means that if you need to make sustainability a special case to get it pass the system, there is help from the legislation. But go a bit further, make sustainability part of everyone's regular business practice, not just the specialists' – make it part of everyone's day by inviting them to participate. Don't spend too much time talking, debating or setting up committees – just get the leadership to push on with it.

Both the sustainability specialist and the Egan advocate can use legislation to compel a degree of support and by using the ideas recommended in this book, help the leadership to become more committed and advancing both the Egan agenda as well as doing sustainability.

NOW: DRIVER ON LEADERSHIP

Corporate commitment to sustainability improvements and the company taking a pragmatic and professional approach to making this happen using legislation and social pressures as the necessary leverage.

2. Focus on the Customer

In 1998, society has little say in construction efficiency, but nowadays social pressures can shape the policy on sustainability. The elements in Rethinking Construction Report: improved efficiency; commitment to people; focusing on the customer will all resonate with a society that is more aware of its influence and supportive of sustainability. Developers, designers, builders and building managers are all aware of society's demand that we become more sustainable in our operations. Problems of how poor design hampered building users get highlighted quickly[11] and opinions will start to be shaped by user demands as well as client needs.

The past decade also brought in the concept of corporate responsibility. While this has limited impact on the actual building methods, the pervasive acceptance of corporate responsibility means that the construction sector has started to pay a bit more attention to the end-user.

Because of its whole life nature, sustainability is helping the construction project team to rediscover the other components besides their own specialisms. The legislative push for tracking the way materials flowed on and off the site also encourages a broader focus.

The entire value stream (including the supply chain value stream) needs to be aware of how it can be made more sustainable. The customer is the user of your products and services, the client is the one who pays for your services and products and the end-user are people who uses what you build. These three groups are intertwined and linked. The driver is the end-user who expresses a need for sustainability in buildings. This is then interpreted by the developers and passed on to the designers.

11 See footnote 12 in Chapter 10.

NOW: DRIVER ON CUSTOMER FOCUS

An integrated value stream paying attention to both the big picture and the immediate elements.

3. Integrate the Process and the Team Around the Product

The Rethinking Construction Report was very direct on this: the most successful enterprises do not fragment their operations. However true this is, it is difficult to apply this in construction because our industry, unlike others, decided that our more frequent boom and bust cycles are best endured by a fragmented subcontracting infrastructure. For better or for worst, this is our industry structure and we need to live with this reality and make the most of it. The global car industry consolidated to major car makers and thence power is concentrated. The UK construction sector is not like that and unlikely to go that way. While our industry may never reach the levels of efficiency in the top car factories, we can do something is make sure the individual fragments are as efficient as possible and that handovers between the fragments are managed effectively.

However, our fragmented industry also means there are pockets of fresh thinking and innovation everywhere. We need to celebrate the diversity and not always clamour for commonality and conformity because those lead to dull, unimaginative and uninspiring outcomes, and sustainability in construction is not like that. The diversity will provide a range of solutions to meet the needs of diverse end-users.

What is important is to demonstrate openness and respect for other ways of doing things and also take into account that:

- Whatever you designed, someone has to be able to build it within a reasonable cost and users have to be able to derive value from it.
- Whatever you built, it needs to be true to the vision of the design such that the end-users gets to use/live/enjoy the values created by the designer.

- In other words, work together to solve problems because sustainability is not something to do in isolation – it belongs to everyone.

- Lowest price bids are likely to deliver lowest rpice solutions – not necessarily the best choice with eco disasters around the corner and social pressures are pushing us for better solutions. Partnership is the way forward for sustainability.

- And most importantly, move away from building sustainable designs using traditional (or biblical) techniques and instead build sustainable designs in a sustainable way.

NOW: DRIVER ON INTEGRATION

Use the diversity in our industry to bring about sustainability in many different ways through many different partnerships. Even though the cost of building is low over its useful lifetime, both in energy terms and monetary terms, it is not a reason to ignore the waste in embodied energy lost in the skip loads of materials we waste. It is every team on site working together to cut material wastes.

4. A Quality-Driven Agenda

This is valid both for 1998 and nowadays. Customers never get tired of quality except when it is time to pay. But once again, people are coming round to understand that sustainability, like quality, is not going to be free – low carbon technologies will cost money. However, a quality-driven agenda means driving out elements that causes variance in quality. In plain English, it means driving out waste.[12] From what I explored in this book, this also means doing sustainability. It does not often get as simple as this.

Egan acknowledged that with a 'lowest bidder wins' mentality, it is difficult to deliver an industry that is high quality. This is equally valid for sustainability, although legislation does help to set up some minimum level of performance. Whether this will satisfy customers is another matter. Although Egan also suggested exceeding customer expectations, I think some caution is

12 A small caveat here, it will drive out waste only if people do something about the variances in quality.

needed when it comes to sustainability. BATNEEC still needs to be taken into account. Nevertheless, what is not wanted is the tick-box mentality that some are applying to the Codes for Sustainable Homes where poor quality concepts and approaches are adopted because they are the cheapest way of meeting the Codes.

Quality is not free, neither is sustainability. However, a quality-driven value stream will deliver the best value (the real thing, not the government's jargon version) for sustainability because it will look at the causes of poor quality or waste. Cut the actual wastes, cut the causes of wastes and you are doing sustainability.

NOW: DRIVER ON QUALITY

A quality culture helps bring about sustainability.[13] Quality is not necessarily free, but improved processes can save money now wasted through poor organisation and messy workplaces.

5. A Commitment to People

The Rethinking Construction Report stated that a commitment to people goes beyond decent site conditions, fair wages and care for the health and safety of the workforce. It should also include training and development to make the most of the workforce's talent. This type of thinking is quite prevalent in the car industry in 1998, especially with the Japanese producers who invented Lean operations. Egan again acknowledged that the fragmented nature of our industry made this difficult. Yet the Report was really saying a lot of the things that many companies now include in their corporate responsibility statements. It has become social sustainability because it is the right thing to do.

However, we are not there yet. Although the trade press showed the best construction companies compare well with top employers in other sectors,[14] we still have problems with decent site conditions, fair wages and care for the health and safety of the workforce. In 2007, I read[15] that several workers died

13 The well-known quality mantra: 'get it right, first time, every time' is also a mantra for sustainability: doing this means optimised costs, motivated people and minimised impacts on the environment – what's not to like?

14 *Building Magazine*, Employers Guide, October 2008.

15 *Construction News*, 11 January 2007.

while sleeping overnight inside a freight container on site. In 2009, I read[16] that a site employer was jailed for three years for having 'a laissez-fair attitude to health and safety' which led to the death of a 15-year-old construction worker.

We just do not respect the people we hire to deliver results for us. I have had supervisors telling me it is pointless trying to raise the awareness of their operatives – 'just tell them what to do'. I have had training managers from global construction companies telling me that they want a sustainability programme that tells their people what to do,[17] their people just need to know how to separate waste not why the waste occurred in the first place. Derek Stow of Measom, a dry lining contractor, had a simple answer for people who think like that: 'fire them'.

People are your best assets. True in 1998, and still true now – people are more than your best assets, they are the only assets that can make things happen. Without the right people, nothing gets built. It does not matter whether the right people are migrants or native workers, what matter is that they can do the job in a sustainable way and that means the right training to develop the green capabilities, discussed in Chapter 5.

NOW: DRIVER ON COMMITMENT TO PEOPLE

People are still your best assets – respect, recognise and reward their efforts and they will help you to do sustainability. Respect the people doing the actual building work and invite them to come up with better ways to do the building.

Applying Egan's Vision

Despite the reluctance for change and the conservative nature of some parts of the construction industry, Lean operations is the way forward for any business, and the sooner our industry embraces this wholeheartedly, the sooner our quality will go up and our costs come down. It will also let you get started on sustainability. However, unless you are making cars in Toyota's factories, the Toyota/Lean approach needs to be adapted instead of being adopted wholesale. Although construction and car making have many similarities, they also have many differences.

16 *Construction News*, 20 July 2009.
17 Needless to say, I did not win that contract!

The similarities are that both the car and construction industries are aiming for lowest costs. We in the UK moan about the culture of lowest bid being the winner. But the car industry is equally cut-throat and low cost is the mantra there too. Squeezing the subcontractors is something we are not too proud to discuss openly in the construction sector but it is equally tough in car making and they squeeze just as hard.[18] Nevertheless, Lean is probably the best management approach to lowering costs without losing competitiveness.

One major difference, however, is that in the car making world, partnerships with suppliers are the norm. Egan was unhappy about the level of integrated working in construction, so is BIS on the Strategy for Sustainable Construction.[19] Another major difference is that in UK construction, much of the work is subcontracted whereas in car making, significant part of the value-adding is done by in-house staff. This type of employment pattern means that workforce development is a worthwhile investment (since they are your employees) and the workers can then see how the improvements they come up with make their own jobs more interesting. Frameworks, a solution to the lowest bid wins mentality was moderately popular until the 2008–09 recession when companies started to revert to lowest bid approach. Nevertheless, to some smaller companies, getting onto a framework can appear like expensive pre-qualification plus a 'beauty contest' rather than an opportunity to work together and innovate.

Consistency and Quality

The Report bemoaned the fragmented nature of our industry and how this hampered getting lower costs and also more consistent outputs. Before going to consolidate all the thousands of construction companies, perhaps we should start with the Major Contractors on consistency. Many of the Major Contractors have multiple projects going on at the same time. However, often every one of these construction project is like a 'franchise' or an independent subsidiary. It has the corporate logo and some corporate trappings, but it is also the project director's fief.[20] There is often a stronger imprint from the local management team rather than from the head office. The different fiefs mean a loss of

18 A few years ago, I worked with a small car parts supplier who was told by their customers to deliver a 10 per cent year-on-year price reduction. Egan also advocated something like this for the construction industry.

19 The Strategy for Sustainable Construction 2008, available from the BIS website.

20 This is equally valid for subcontractors – some perform better, others less so, regardless of their company's reputation.

consistency because although teams for different projects will build roughly in the same way, the individual site cultures will be different. For a large construction company, such as one of the Major Contractors, there will be a very large number of independent subsidiaries that carry the same badge but do things somewhat differently. This is no way to run a corporation because it does not make sense, and it does not enable a consistent sustainability programme. Clients and regulators can find working with such a Major Contractor difficult since every project will do the same things slightly differently with somewhat different outcomes.

However, this situation of many franchises is quite similar to another type of businesses – the multi-shop service chains and franchises where the personality and leadership of the local managers often make the difference. The construction industry is quite familiar with this type of business (and not just because we built their buildings) but because many of our teams spend their lunchtimes there. Yes – it is the fast food chains.

A successful fast food chain works despite often having staff turnover of over 100 per cent each year. They do it with people that have the right capability to flip burgers or work the tills. Sustainability in construction also needs people with the right capability. A successful example of large staff turnover at multiple locations is the McDonalds model. These eateries are based on consistency of performance at every location, on every occasion. This is also what we want for our industry – a consistent performance across all projects generally and specifically for sustainability.

How does the McDonalds model work? By sticking rigidly to agreed procedure and processes, and using similar equipment coupled to consistent basic training, you end up with the same menu, decor, taste and flavour – in other words, consistency. With consistency in place, quality and waste can be tackled systematically and skills build-up accordingly.

For sustainability to be realised, every construction project needs to be consistent in its operations. This can be achieved by rigidly adhering to an agreed approach to enable consistency. This means very well-designed Method Statements, not the Method Statements that some manager wrote a few years ago for a different project, but a conscious effort by the company to research the best practice (not entailing excessive costs) and these become the regularly updated Method Statements. Subcontractors are encouraged to adhere to these but also to improve them through a formalised process.

There will be howls of complaint about every project being different and we are not selling hamburgers. Much of this is likely to be a resistance against change. People are comfortable in doing what they have always done. But for the company there is no choice: business survival is not guaranteed in our economic system and sustainability means successful change, evolution and adaptation. Both the Core Intent statement and the House of Sustainability tool will allow you meet local requirements (and not just the project director's whims) while remaining within the corporate perimeter.

For many colleagues in construction, suggesting McDonalds as a model may be too much to swallow, so let us examine other situations where staff are rotated to form *ad hoc* teams regularly without losing much in corporate consistency.

An airline sorts its crews out by rota and so you may fly with a crew that has not flow together before. However, after a pre-flight meeting, everyone knows their roles and how these roles interact with the other crew members. Again, focused training and adherence to working methods mean that different teams can deliver a consistent service to the passengers on a journey. You may think that a flight is like any other flight (while every building is, of course, different), but every flight is different: a different assortment of awkward customers; and different weather conditions, for example. The crew deals with this through discipline, initiative and training within the parameters of their working methods.

Now I have upset even more colleagues by suggesting we benchmark airlines (carbon footprint and offsetting not withstanding), perhaps we can explore another model of *ad hoc* team. In a hospital, the team doing operations are also based on rotas. They work together to open you up and sew you back together seamlessly (well, maybe not seamlessly in reality, but you know what I mean). Again it is professionalism and teamwork based on working methods and discipline.

Our industry is no different. We work with an *ad hoc* group of professionals and get the job done with a large number of people that we may never see again. To deliver sustainability (and consistency), we need to stick to agreed and up-to-date working methods. My suggestion is for every project by the same company to share the same top-level sustainability objective and yet retain its local objectives. All the projects will have a workplace organisation scheme that deploys some form of workplace effectiveness like 5S. All projects will invite participation from subcontractors to reduce waste and innovate – not to appropriate blame, and all suggestions will be received with respect, efforts recognised and rewarded.

When an operative comes onto that company's project site, he or she will know that things are done in a specific way there: it is a good company that respects its workers and does things properly according to the Method Statements that are regularly updated and make sense.

Time is Not on Our Side

It took Toyota nearly 20 years to get their just-in-time system sorted. A lot of these improvement activities will take longer to become embedded in a company than the average time of a construction project. The only way to make this happen in a construction company is to make sustainability a practical part of the culture and a corporate directive for every project. That means sustainability is not just the posters and slogans, but actual practices and that means an unwavering corporate commitment.

What will happen after the slow recovery? Companies that are Lean and ready will win the most and stay on top. Or will the next winners be the people who make construction an 'assembly' business by using off-site fabrication? Where will your company be? If you are still using out of date business practices, will your company be sustainable? You cannot rely on Sir John Egan to give you another wake-up call in ten years' time, because it may be too late.

A Commitment for Sustainability in Construction

Sustainability in construction needs to be more pragmatic. Not pragmatic in terms of doing the minimum and hoping that regulators, shareholders and society will not notice, but pragmatic in accepting the world has changed and the way we do business has also to change. Our industry is well know for its pragmatism.

Every practice we gloat or moan about is based on one or more element of pragmatism:

- Too many boom/bust cycles? Rely on subcontractors.

- Our people's skill not up to scratch? Hire skilled migrant workers
- Our plans a bit suspect? Order 10 per cent more materials as a buffer.

Laugh if you wish, but despite these attitudes, we have managed to build the infrastructure of this country. What we need now is our pragmatic leadership to focus on making sustainability a part of the way we do business.

We need to confront our suppliers and our clients about their commitment to sustainability too.

It is no longer about asking them to put their money where their mouth is. Sustainability is no longer a debating point, money has to be spent because sustainability is here to stay. This will be part of our working lives as well as for those who come after us. Our clients, including the government, have to accept that we need to work together and not play the hypocritical game of saying one thing and doing another. Neither the industry, nor its clients or its supplies can get away with that sort of behaviour in an age when anyone can prepare a report lambasting our performance and the whole world can read it within 30 minutes.

Our industry has been accused of taking a short-termed outlook, but can we really define a company's horizons in such strict terms when industry-changing events are happening so quickly?

One pragmatic way of looking at this is simple: in the shorter term, make sure you stay profitable; in the longer term, transform the company to a sustainable footing:

- Optimise profitability in the shorter term because otherwise there will be no long term.

  Action: sort out the financial obligations.

- Carry out workplace organisation to cut waste and give yourself some breathing room.

  Action: save money by cutting waste.

- Then optimise the effectiveness of all operations by improving the way business processes are carried out.

  Action: build in a sustainable way.

- Plan for the future by sorting out the present and getting your people ready to take on whatever the future brings.

  Action: respect your people whether they are subbies or migrant workers.

Sustainability needs a commitment to pay for the necessary improvements, infrastructure and capability development. Whatever the 'evidence' shows, moving from a business-as-before model to a sustainable business model will cost money. People need to have their awareness raised and you will need to pay: someone to write the awareness raising contents; then you will need a folder to hold the content; then a filing cabinet to store the folder; and staff time to do the PR about your awareness raising content. All this costs money and it is not stuff you can offset by buying a forest somewhere else. If you hire a sustainability specialist, there is the total cost of employment to consider (salary, pensions, expenses, desk space, telephone bills, etc.) and since these people are in the value-enabling part of the work your company does, they are a physical cost. Transforming the organisation also brings with it a switching cost.

Sustainability needs a commitment towards continuous improvement. Why? Because the only way we can afford sustainability in this transitional era when not everyone accepts that sustainability is the way things are done, is by cutting waste. Otherwise, the journey will be painful and less successful because we either cannot afford it or will achieve less with a stingy budget. Cutting waste means continuous improvement means saving money.

Skills and Skills Development

We also need a change in our vision about people, capability and skills. We need a commitment towards the reality behind this exhausting (and seemingly endless) debate.

The reality is that people in the UK are not entering our industry as apprentices because the pay is lousy, the conditions are tough and not enough companies want to bother with apprentices for all kinds of reasons. The result is an industry that lacks the necessary skills, while what we needed is, according to David Vincent of the Carbon Trust, a government agency on sustainability, is 'the creation of a cadre of practitioners at all levels in the construction industry, with skills and a growing bank of experience in the design, construction and operation of sustainable buildings'.

So what do we do? Going to a McDonalds model sounds like de-skilling. Going towards off-site fabrication sounds like de-skilling. Going for a disciplined-style of strictly following working methods sounds dull and will turn people into robots, and yes, it will de-skill our people. Does it matter if the UK construction industry lose its 'skills'?

Construction in the UK has been supported by cheap migrant labour and now with the slow recovery, wages will remain suppressed. This can mean that we start at a lower base when the recovery really gets going. And there may still be some Eastern European countries that can supply low-cost skilled workers when we need it.

This is actually quite a viable strategy for the industry. We do not have to train people: instead we just hire people trained by other countries because we can afford to pay a little more. This model will work as long as we can afford to pay 'a little more' to attract other nationals to work in the UK. It is our industry and we shape it the way we want to, not because some politicians wants help with their agenda. Is this what I am recommending? Well, this is actually the reality, just like the need to do sustainability is a reality. As long as we are not prepared to pay the wages that will attract local people into our industry, then we need to live with the reality of not having sufficiently trained local workers during a boom period. (Of course, during a bust period we have different things to worry about.)

The situation described is also not de-skilling because it is about different ways of working. The matter is not about how detailed the training manuals are or whether pre-fab assembling skills are less desirable than the biblical trades skills. It is about the outputs: it is about customer satisfaction. If the UK clients are happy about migrant workers, then we deliver with migrant workers. If the UK public likes to live in pods, then we deliver pre-fab pods. At the end of the day, we are not style gurus telling the public or clients what they should want, we just do what the clients ask and pay for.

The important part about the skills debate is that we need to deliver a consistently good customer experience. Before you wince and think here comes the 'marketing speech', I need to tell you that the choice of 'good customer experience' is deliberate. A good customer experience means all the usual contract stuff (price, schedule, quality) and also that the client enjoyed working with you.

I believe it does not matter where we get our workers from, instead, what matters is what do we do with these workers once they join the industry. Training should not only be funded from grants but also from company resources. You can make training part of the share out from a Gain-Share programme. And increasingly, the training needs to include the sustainability capabilities of the workforce: to reduce waste; to be innovative; and to work together to do sustainability.

Ultimately, all these commitment boils down to a commitment to excellence, which is really a commitment to sustainability.

Appendix 1: NoWaste: A Case Study

NoWaste is a sustainable construction programme I developed in 2004 through a Carbon Trust supported initiative with Bovis Lend Lease (now Lend Lease), UCATT the construction union and the London Business Link. Since then it has appeared in various forms with different partners. This case study focuses mainly on the most recent implementation at the 'big sports project at Stratford' (see explanation for 'no names allowed' in Chapter 4) in 2009–10. The NoWaste project continues in 2011 with contractors, such as Carillion and Skanska, at the even bigger construction project at Stratford.

What Is It?

The NoWaste concept is the same as this book's – engaging a majority of the workforce to take part in sustainability activities through reducing waste. The underpinning tools and techniques of NoWaste are Lean operations and Six Sigma Quality with the programme presented through a cartoon interface (see Figure A1.1).

Figure A1.1 NoWaste cartoon

The implementation at the 'big Stratford project' was that of a training programme. My colleague Mike Harris and I implemented it with the support of Andrew Kinsey, the senior sustainability manager at Lend Lease.

The NoWaste elements at Stratford comprised:

- sustainability training (3-hours duration)
- managers and specialist programme (1.5-hours duration)
- NoWaste *Espresso* sessions (30-minutes duration)
- optional NVQ programme.

The sustainability training element is aimed at everyone and is made up of an introduction to sustainability (as seen by the NoWaste programme – so waste reduction takes centre stage); spotting and solving waste problems (with exercise sessions); as well as an element of generic training that is mapped to an NVQ award (see below). This portion of the training is intentionally jargon-free with the Lean/Six Sigma tools are shown simply as problem-solving techniques.

The managers and specialists programme is aimed at explaining the NoWaste concept, themes and techniques in greater detail. As the managers and specialists are assumed to have attended the sustainability training element, the various concepts and techniques are then explained with references to their origins: Triple Bottom Line; Lean operations; Six Sigma; and so on. The aim of this element is to enable managers and specialists to support their teams in the waste reduction activities.

The NoWaste *Espresso* sessions are designed for team-based problem solving. Although the sessions resemble Tool Box Talks (and are indeed marketed as 'supercharged Tool Box Talks'), they are rather different to the traditional 'sign your name, shut up and listen' programmes. The *Espresso* sessions are focused on cutting the Eight Hidden Wastes (see Chapter 9) through spotting, analysing and dealing with waste problems. Although this element is entirely based on Lean and Six Sigma techniques, these are hidden in a specially designed interface that guides the team through a sequence of identification, analysis and action planning.

During these dynamic sessions the team and their supervisor discuss a particular waste problem they faced, spotting and solving problems, for example:

> **Situation**: the team wastes time every morning going through their piles of fittings to get to the components they want for that day. Owing to the layout of the workplace and multiple storage areas, this can take up to 20 minutes each morning.
>
> > *Supervisor*: (pointing to the poster of overstocking waste) We have this every day, why do you think it happened and what can we do about it?
> >
> > *Operative 1*: Because the place is a mess and we over ordered to make sure that if there are delivery delays later, we don't get caught out.
> >
> > *Supervisor*: So is it a case of people over ordering or just messy housekeeping? ABC tripped and fell over last week with all the waste packaging lying around.
> >
> > *Operative 2*: Probably because they just piled everything into the nearest storage area instead of setting it out into the right areas and everyone then rips the boxes apart to get their component and messes it up.
> >
> > *Operative 3*: We got two problems, the delivery and the storage.
> >
> > *Supervisor*: Okay, let's stick with the storage one first because we can do something about it easier. Why do we slap everything into the nearest storage bin?
> >
> > *Operative 2*: Because the labourer has no idea where to put the stuff.
> >
> > *Supervisor*: Why doesn't he know where to store the stuff?
> >
> > *Operative 2*: Because nobody told him.
> >
> > *Operative 1*: No, because every one tells him something different.

Supervisor: Okay so we need to have an agreed method of putting components into the storage bins, how will we do that ...

The *Espresso* sessions can readily take place at the workplace, rather than the training room or project office. The special interface means that any supervisor can guide the team through a waste reduction exercise without extensive or expensive training.

The optional NVQ programme was linked to a Level 2 NVQ award – Removal of Non-Hazardous Waste (Construction). There were several reasons for this particular choice: the offering of a recognised qualification to operatives and supervisors as well as using an NVQ award as a motivating factor. That this particular NVQ award was not necessarily anything to do with sustainability was a minor problem(!), but as it was the nearest available award that has a link to both waste and construction,[1] it was an act of expediency.

The NVQ-linked programme, deployed during the pilot stage in 2009–10, was modified following the changes to NVQ funding regime from the UK government in 2010[2] when it became an optional element.

What We Did

The NoWaste implementation started with a promotional campaign to individual contractors advising them of the benefits of good sustainability practice and differentiating the NoWaste programme from other sustainability or environmental programme ('twice national award winner!' and 'big project at Stratford award winner!'). The marketing period usually took a bit of time with the regular movement of project management staff in a construction project. Once a few contractors signed up, then peer pressure was deployed to encourage the others. The final few laggards were further encouraged with specific contract clauses which stipulated their taking part in environmental training.[3]

1 In discussions with ConstructionSkills, the sector skills body, it became clear that unless there is an occupational role, then CSkills will not create a set of occupational standards. This is probably due to a belief that skills not capabilities (see Chapter 5) is what matters. We therefore enter a situation where 'if there is no occupational role for sustainable carpentry, then we cannot have a set of standards'. Logically, this may make sense, but practically, this is not really an optimal approach.

2 In simple terms, the money ran out.

3 Anecdotal evidence tells us that in construction, even legally mandated health and safety programmes sometimes do not always draw a full audience – the reason is usually 'busy

Welcoming and cajoling, we invited our delegates into the training sessions which ran weekly with a sustainability training session in the morning and a managers and specialists programme in the afternoon.

Espresso sessions were arranged with various contractors and took the place of a regular Tool Box Talk slot. For these, Mike Harris and I usually plan for several sessions in a morning and similarly in the afternoon. Contractor-specific *Espresso* packs were made with their logos (as opposed to the overall project logos).

Did it Work?

A pilot programme ran in the autumn of 2009 to early 2010. This was mainly working with teams carrying ground works as well as some concrete framing. Analysis by Bovis Lend Lease[4] attributed a waste levels reduction of 0.7m^3/£100K spend to the NoWaste programme. (The waste volume per £100K spend is a Key Performance Indicator (KPI) for Bovis Lend Lease.)

The 2010 programme showed[4] that during the 'big Stratford project' fit out stages:

- Average waste KPI for projects where NoWaste was applied 6.78 m^3/£100K spend.
- Average waste KPI for projects where NoWaste was not applied 8.72 m^3/£100K spend.

This was worked out to equate to savings of about £136,000 for waste disposal costs alone.

Feedback from individuals attending the programme was also very positive. Generally, the NoWaste approach was not what they were expecting:

- they were not talked at, or told what to do;
- the contents were relevant to their work;

meeting schedule'.

4 Source: 'big Stratford project', 2010 (sorry about this reference, but I do not make the rules on marketing the big sports event).

- they were invited to participate; and
- they learnt something new about sustainability.

We had quite a lot of enthusiasts at the Stratford project, from managers to labourers. The NVQ was popular too, for some it was their first qualification and a boost to their self-confidence.

Typical positive feedback:

- 'learnt something new';
- 'learnt something useful';
- 'everyone should attend this';
- 'should make our managers attend this'.

What Worked Less Well

To almost all the people trained, NoWaste is an unexpected programme. This means there is a likelihood of not meeting expectations if some of the attendees have preconceived ideas on what a sustainable training programme should be. We therefore receive some feedback along the lines of, 'this is common sense' or 'everyone knows this'. These comments are interesting because if it is common sense or everyone knows this then why are these not embedded into everyone's daily activities?

But there are also some more concerning comments:

- 'operatives don't need to know this ... just tell them what to do';
- 'this is not sustainability';
- 'this goes on for too long'.

The first comment is concerning and quite commonly encountered in businesses where command authority is deemed desirable. But it also limits the development of the workforce and ultimately the productivity. It also assumes knowledgeable and infallible leaders – the former is likely but the latter is not.

'This is not sustainability' – this comment although concerning, is not particularly serious. If the individual does not subscribe to our suggestions, and thereafter do nothing, nothing is lost. The potential gains remain potential and life goes on for the individual. It is his/her loss.

The main negative response from our training workshop survey is about the length. This is a difficult one to address. The sustainability training programme lasts about three hours, which is a compromise because it is unlikely that contractors will allow their teams to be released for anything longer than that.[5] Although a whole day or two days will allow much better retention of the information as well as a better grasp of using the tools, the reality of the construction site means that the contents will need to be tailored to both the needs and the constraints.

It is my belief that the complaints about lengths are due more to boredom and work pressures than the actual length of the programme itself.

The boredom issue was addressed readily. When I coupled the NoWaste programme to the NVQ closely, it *was* dull. I cut out much of the (very dull) details from the NVQ and reworded the details such that nobody fell asleep. Construction is an active industry and most people are less fond of sitting around than doing something. We tinkered with the programme: shorten our talk; revamped the NVQ elements; and extended the group work sessions on using the waste reduction techniques.

The work pressure issue we can do little about because just like many other projects, there was always a 'rush' on.

We invited a researcher, Martina Schniepp,[6] to review and assess our programme, and she reported that the main issue she found was that sometimes the workshop participants were very quiet and not too active. Mike and I tweaked the programme to add in extra interactive elements (such as the Common Knowledge in Sustainability Quiz – see the Appendix 2).

5 This is not unique to construction, but also happens in factories. Hourly paid staff generally tended to get their training in half to one day.

6 Martina Schniepp, MBA Dissertation, London South Bank University, 2010.

A more difficult problem we faced was the lack of a follow-up mechanism. As NoWaste is implemented as a training programme, follow-up became more difficult with teams on different site plots.

This was partly because the big project at Stratford was originally managed by Lend Lease and the aim was to build the necessary athletes accommodation, and then sell them after the 2012 Games to recover the investment. However, with the recession/credit crunch, Lend Lease had to turn to the UK government to fund the construction and this resulted in a somewhat complex management situation.

Some of the plots remain under Lend Lease's direct management while some others were opened to bids from other contractors. This meant that Bovis Lend Lease[7] was the overall project manager for every plot as well as the main contractor on specific plots, alongside other Tier 1 contractors, such as Ardmore, P. Elliott[8] and Willmott Dixon, who manage their individual plots. As such, a classical implementation of NoWaste is likely to resemble a logistical nightmare. It was therefore agreed that NoWaste will be presented as a sustainable construction training programme instead.

Follow-up instead came from two sources:

- Key performance indicators from Bovis Lend Lease – the volume of waste generated for every £100k spend.
- Informal reporting from NoWaste attendees (managers, supervisors, skilled trades and labourers) – these included anecdotes, photographs as well as discussions and impromptu problem-solving sessions.

Support From the Specialists

The NoWaste project tended to have a fairly consistent response from sustainability and environmental managers at the 'big Stratford project'. The pattern is a two-step:

7 The company dropped 'Bovis' from its name on 2011.
8 P. Elliott went into administration in 2011 before completing its plot, see the trade press for more details.

1. initial (mild) scepticism, followed by

2. enthusiastic support.

The initial response is not surprising as NoWaste does not have environmental management roots but instead has an industrial parentage. As such it does not figure in 'traditional' sustainability management thinking. However, once they became more familiar with the programme (or 'got brainwashed', as one of them mentioned to me), they became more supportive. I believe it is because of the following reasons:

- NoWaste does not compete with their own programmes. While this may sound a bit ungracious, nobody likes people messing around on their turf.

- NoWaste can make them look good. It is a programme that can be deployed readily and has a suitably impressive pedigree.

- NoWaste complements their activities. By focusing on the operational elements rather than strict compliance, it gave some degree of ownership to the non-specialist construction worker, whose contribution ultimately adds to the compliance requirements.

Analysis

If the NoWaste programme is analysed using the various management models described in this book, it is possible to see a few general themes.

The Balanced Scorecard (see Chapter 3)

NoWaste readily and positively slots into the Balanced Scorecard regime of:

Financial perspective: it helps save money; release money from cash flow; and increases cash in hand. In fact, it recovers money deemed lost through an acceptance of wasteful operations.

Customer/client perspective: reducing waste; reducing cost; engaging people; innovating – these are all activities that pleases the customers or clients. Adding to these is the possibility of good PR and general goodwill.

Processes and procedures perspective: by innovating and improving the way work is carried out, processes and procedures are enhanced.

Developmental needs of staff perspective: NoWaste brings a capability to 'do sustainability' as a lifelong skill.

The Triple Bottom Line (see Chapter 1)

By marking waste as the anti-sustainability, NoWaste offers a simple and direct way in which everyone can 'do sustainability' and thence reduce costs, get involved and reduce environmental impacts.

The Kano Model of Customer Satisfaction (see Chapter 12)

NoWaste fits well into the 'unexpected' category as it is a new approach (to most people) about doing sustainability:

- it is aimed at everyone, not just managers and supervisors or specialists;
- the contents are pitched at a working construction site, not a theoretical analysis;
- the programme does not 'talk down' to the people attending; and
- it invites opinions and suggestions from the most senior to the most junior members on site.

This provides the excitement and offers information on sustainability in:

- a new context where everyone participates; and
- new contents aimed at the way work is carried out in reality on construction sites.

Although the concepts are not new, the two 'new' elements above are sufficient to get people excited.

Acknowledgement

The NoWaste programme at the two 'Stratford' projects were partly supported by ConstructionSkills' Management and Supervisory Development Programme (thank you CSkills!).

Last Word

NoWaste (the project) and *No Waste* (the book) – sorry, imagination ran out and we decided that the title should be true to the contents of the book, so *No Waste* it is.

Appendix 2: CKiS – Common Knowledge in Sustainability

Common Knowledge in Sustainability (CKiS) is a simple tool that can be included in any training, or Tool Box Talk sessions. It is a series of simple questions about sustainability that is aimed at focusing the minds of the teams being trained (see Figure A2.1).

It was developed during our 'big Stratford project' following encouragement from ConstructionSkills. As such it takes the form of CSCS touch screen tests. I included a collection of 30 questions for the book.

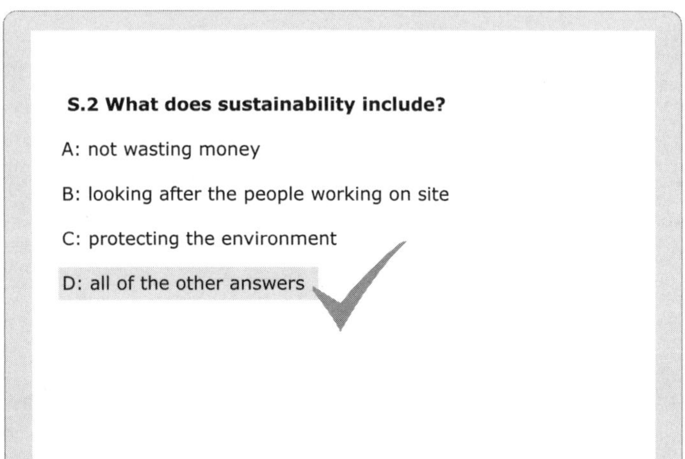

CKiS
Common Knowledge in Sustainability

Figure A2.1 Common knowledge in sustainability

I found during piloting that information retention is quite good if the information related to the questions is integrated into the training programme. Generally, out of say 20 questions, most people, whether managers, supervisors or operatives were getting about 16–18 correct answers.

The initial draft version had some ambiguous questions and answers, and instead of clear-cut answers we often ended up with an interesting debate on how to deal with various sustainability issues. Besides having to redraft these questions, I also realised that when given both the right encouragement and the right context, our industry has a workforce that is capable of making serious progress on sustainability.

The questions below cover sustainability generally, with a few about the various pieces of legislation, some on water-related issues and a few on the waste labelling. There are also questions on waste reduction, reuse of waste materials and the sustainability benefits of working to standardised Method Statements.

The CKiS questions, when matched with a suitable training programme, can help construction workers understand:

- What the various bits of legislation mean in their context.
- How construction materials are made (and how much energy is used in their production).
- That they have a role in bringing about sustainability in construction.
- The type of situations they may encounter at work and how to deal with them.

Deploying CKiS

I found that showing the question on screen in a 'touch screen test' lookalike slide presentation quite effective as the team being trained called out the answer (and also to colleagues whom they reckon got the wrong answer). Generally I only show one question at a time and spread these across the entire training session, thus brightened up the training programme and also keeping the attention of the group.

CKiS is developed with assistance from Andrew Kinsey of Bovis Lend Lease and my colleague, Mike Harris.

The Questions (a Powerpoint Version is Available on the CD)

Note: I decided to keep some of the questions ambiguous to stimulate discussion. The 'correct answer' is highlighted by an *. In some case, several of the possible answers are correct but as this quiz is about sustainability, some of the otherwise correct answer are not selected.

1. What can you do to for sustainability on site?

A: work safely

B: segregate waste

C: turn off engines and motors when not needed

D: all of the other answers*

2. What is sustainable development?

A: development that sorts out our needs now

B: development that both deal with our current needs and also our children's future need*

C: development only about the future needs of our children

D: development that is concerned with giving aid to foreign countries

3. Before you put something in the skip, what should you do first?

A: make sure there is a label on the waste

B: tell your supervisor

C: check whether someone else can make use of it*

D: make sure there is room in the skip

4. When a product has this label, how should you dispose of it? (✖ on orange background)

A: put it in the nearest skip or bin

B: check the product label for disposal instructions and follow them*

C: ask your supervisor

D: leave it somewhere for other people to deal with

5. Keeping the workplace clear and tidy can:

A: help reduce slips, trips and falls and make it safer to work*

B: make the site look better and impress the client

C: cut waste removal costs

D: make the local community like us better

6. What does sustainability mean for you on site?

A: obey environmental and health and safety laws

B: minimise damages to the environment

C: reduce wastes

D: all the other answers*

7. The benefits of working to standardised Methods Statements are:

A: we will not get told off

B: it is safer

C: it reduces costs

D: it reduces wastes

E: A, B and D above

F: B, C and D above*

8. How do we best save water on site?

A: not using water to keep down dust

B: not hosing down vehicles wheels when they leave the site

C: not leave taps running after we finish our work*

D: not hosing down our boots

9. Who is best at spotting waste on site?

A: the architect

B: the client

C: the site manager

D: everyone on site*

10. What is the Landfill Tax?

A: it is a tax on construction generally

B: it is a tax on buying land to build on

C: it is a tax on disposing waste*

D: it is a tax to help local authorities to earn income

11. Why should we cut waste?

A: because construction materials take a lot of energy to produce*

B: because construction materials are heavy

C: because a lot of construction materials can be reused

D: because construction materials take up lots of storage space

12. What is the Code for Sustainable Homes?

A: a new building code for safety

B: rules on building homes for a sustainable lifestyle*

C: an advert for developers to charge more for their houses

D: a new TV reality show

13. How is cement made?

A: dug out of the ground

B: crushing and roasting rock and minerals*

C: an oil by-product

D: melting sand

14. Which of the following causes waste?

A: over ordering

B: poor storage

C: bad weather

D: not working to plan

E: all the other answers*

15. What does the green colour waste label on a skip mean?

A: the skip is for inert wastes

B: the skip is for plaster wastes

C: the skip is for wood and timber wastes*

D: all of the other answers

16. Why do we need to separate plaster and plasterboard wastes?

A: because they are worth money

B: because they can react with other landfill wastes and produce a toxic gas that smells of rotten eggs – hydrogen sulphide*

C: because we are running out of plasterboard and need to reuse them

D: because the client wanted us to do so

17. What does a white colour waste label on a skip mean?

A: the skip is for plastic wastes

B: the skip is for inert wastes

C: the skip is for plaster wastes*

D: the skip is for packaging wastes

18. What do you do with rainwater in a bund?

A: leave it alone and let it evaporate

B: have a drain hole in the bund wall and let it drain out

C: pump it out and dispose as hazardous waste*

D: leave it for someone else to worry about

19. What does an orange colour waste label on a skip mean?

A: the skip is for hazardous wastes*

B: the skip is for plasterboard wastes

C: the skip is for mixed wastes

D: the skip is for metal wastes

20. Do you break the law if you release polluting elements into a watercourse?

A: yes*

B: no, if it is a genuine accident

C: no, if there is flooding on site

D: no, if nobody saw you

21. Which of the following are part of sustainable construction?

A: selecting suitable materials

B: preventing water and soil pollution

C: saving energy

D: all the other answers*

22. What does sustainability include?

A: not wasting money

B: looking after the people working on site

C: protecting the environment

D: all of the other answers*

23. Why should we separate our wastes on site?

A: so it is neater on site

B: so it can be recycled easier*

C: so the client can check what is being thrown away

D: so the local council can charge us a fair amount of Landfill Tax

24. What is Duty of Care?

A: a special duty when working with asbestos

B: a set of legal duties in the way we work*

C: a requirement to make sure we do not make too much noise on site

D: a local authorities requirement to stop us fly tipping

25. Who is responsible for sustainability on site?

A: the client

B: the site manager

C: the supervisor

D: everyone on site*

26. What are ways to show we comply with the Site Waste Management Plans Regulations?

A: discuss wastes with our supervisor

B: record how waste materials are removed by collecting transfer notes

C: separate wastes into different types of materials

D: work safely

E: A and D above

F: A, B and C above*

27. What does FSC mean?

A: it is the Food Safety Council – about the food served in the canteens

B: it is the Forest Stewardship Council – making sure the wood we use is from sustainable sources*

C: it is a stamp for Factory Standards Charter – a quality mark for timber products

D: it stands for Foreign Substance Charter – advising us that the timber is imported into the UK

28. Why should we try to reuse materials?

A: because it saves the client money

B: because it takes lots of energy to make most construction materials*

C: because it makes less mess on site

D: because it is an EU law

29. What does the blue colour waste label on a skip mean?

A: the skip is for metal wastes*

B: the skip is for mixed wastes

C: the skip is for packaging wastes

D: none of the other answers

30. Which of the following are often wasted on a construction site?

A: water

B: construction materials

C: time

D: energy

E: all the other answers*

Appendix 3: The Sustainability Buzzword Generator

Communications is an important part of managing sustainability with the Core Intent statement being a crucial part in getting people engaged. But what about communicating the actual sustainability initiatives? We are often faced with worthy but unexciting project names like Project Build or Project Sus'd.[1] Sometimes, I think we take this all too seriously. (Are you kidding? This is a serious matter! Global warming, climate change – all very serious issues.) Just look at the Al Gore's movie or the Stern Review – neither are a barrel of fun and both take themselves very seriously.

The problem is that neither Al Gore and Lord Stern are doing the day-to-day sustainability management we mere mortals do (they probably have 'staff' to handle those things). To get through the day, we need to lighten up a bit – we also need to be able to poke fun at ourselves too. Otherwise we turn into zealots and then we will lose some of our audience. Well, the Sustainability Buzzword Generator is something to poke fun at ourselves and have a light moment. You can create up to 9,261 unique buzzwords by just selecting a word from each column and immediately make a special contribution to the corporate sustainability initiative!

1 Before anyone feels I am being rude to some overburdened project managers, those two project titles were from my projects. For those overburdened project managers scarcely have time to eat lunch let alone dreaming up fancy project titles, the buzzword generator may make a small contribution to their lives!

Column 1	Column 2	Column 3
sustainable	regenerating	agenda
green	management	initiative
responsible	development	concept
integrated	organisational	programme
equitable	team	project
balanced	policy	capability
future	adaptive	options
renewable	change	plan
ethical	environmental	strategy
social	corporate	forum
holistic	long-term	consultation
stakeholder	responsibility	vision
radical	eco	assurance
fundamental	engagement	matrix
closed-loop	rapport	approach
competency	assessment	network
triple bottom line	convergence	method
total	framework	theory
inclusive	impact	aspect
CSR	ubiquity	operations
global	lifecycle	festival

The Sustainability Buzzword Generator first appeared as part of the 2008–09 Christmas feature at *Building Magazine*'s website. Thereafter it was spotted at various other internet locations. Judging from the comments on the internet, many others also share the moment of lightness. Of course the idea is not new, there had been management buzzword generators galore.

An interactive version is available on the CD, or just search for 'Sustainability Buzzword Generator' on your preferred web search engine for the *Building Magazine* version. Many thanks to Andrew Kinsey of Lend Lease for coming up with some of the words.

Right – time's up! We need to take ourselves seriously again and return to managing sustainability in construction. But do make a note: there is perhaps an opportunity to launch a closed-loop convergence consultation next week, or is it the responsible adaptive forum we are thinking about? Stop grinning, this is a serious matter!

Index

Page numbers in *italics* refer to figures and tables. A page number followed by *f* refers to a footnote.